Klappstühle

Herausgegeben von Werner Blaser

Folding Chairs

Edited by Werner Blaser

Eine Darstellung der Meisterklasse
für Innenarchitektur und Industrieentwurf
Prof. Johannes Spalt, der Hochschule
für angewandte Kunst in Wien.

A presentation by the master class
for interior architecture and industrial design
of Prof. Johannes Spalt, University
of Applied Arts in Vienna.

Birkhäuser Verlag
Basel · Boston · Stuttgart

Eine Publikation mit Unterstützung von Franz Wittmann, Etsdorf-Österreich.

Allen am ‚Möbel' Interessierten wird diese Zusammenfassung von einigen typischen Klappstühlen eine Information über mechanische Möbel geben. Die Firma Wittmann hat sich auf dem Gebiet des mechanischen Möbels über 2 Jahrzehnte erfolgreich betätigt und glaubt daher, dass diese von ihr geförderte Publikation viele Anregungen geben wird.

A book published with the assistance of Franz Wittmann, Etsdorf/Austria

This summary of some typical folding chairs will provide all those interested in 'furniture' with information on mechanical furniture. The firm of Wittmann has been working in the field of mechanical furniture for more than two decades and therefore believes that this book, published with its assistance, will prove fertile in suggestions.

CIP-Kurztitelaufnahme der Deutschen Bibliothek

Klappstühle: e. Darst. d. Meisterklasse für Innenarchitektur u. Industrieentwurf, Johannes Spalt, d. Hochsch. für Angewandte Kunst in Wien = Folding chairs/ Hrsg. Werner Blaser. – Basel; Boston; Stuttgart: Birkhäuser, 1982.
ISBN 3-7643-1357-9
NE: Blaser, Werner [Hrsg.]; Hochschule für Angewandte Kunst (Wien) / Meisterklasse für Innenarchitektur; PT

Die vorliegende Publikation ist urheberrechtlich geschützt. Alle Rechte vorbehalten. Kein Teil dieses Buches darf ohne schriftliche Genehmigung des Verlages in irgendeiner Form – durch Fotokopie, Mikrofilm oder andere Verfahren – reproduziert oder in eine von Maschinen, insbesondere Datenverarbeitungsanlagen, verwendbare Sprache übertragen werden.

Library of Congress Cataloging in Publication Data

Main entry under title:
Klappstühle.
  1. Folding chairs.   2. Chair design.   I. Spalt, Johannes.   II. Blaser, Werner, 1924–
III. Hochschule für Angewandte Kunst (Vienna, Austria).   IV. Title: Folding chairs.
NK2715.K62   1982   749'.32   82-17877
ISBN 3-7643-1357-9

© 1982 Birkhäuser Verlag Basel
English version: D. Q. Stephenson
Layout: Albert Gomm, Werner Blaser
ISBN 3-7643-1357-9

Inhalt

4–11	Werner Blaser Beispiel Klappstühle – eine Bildfolge
12–18	Johannes Spalt Klapp- oder Faltmöbel
19–21	Johannes Spalt Aufgabenstellung
22	Stühle im zusammengeklappten Zustand

23–42	Hocker	
23–25	Damenstockhocker	1
26–28	Japanischer Hocker	2
29–30	Jagdstockerl	3
31–33	Holzstockerl	4
34–36	Feldstockerl	5
37–39	Tragstockerl	6
40–42	Feldstockerl	7
43–53	Kinderstühle	
43–46	Kinderhockerl	8
47–49	Kinderklappstuhl	9
50–53	Kinderklappstuhl	10
54–77	Stühle	
54–56	Gartensessel	11
57–61	Schiffsessel	12
62–64	Schiffsessel	13
65–68	Englischer Klappsessel	14
69–71	Kaminsessel	15
72–74	Holzsessel	16
75–77	Klappstuhl ,Plia'	17
78–89	Armlehnstühle	
78–80	Damenschaukelstuhl	18
81–83	Verstellbarer Klappstuhl	19
84–86	Klapparmstuhl	20
87–89	Regiesessel	21
90–92	Bodensitz	22
93	Nachwort	
94	Verzeichnis der Studenten	
95	Literaturverzeichnis	

Contents

5–11	Werner Blaser Example Folding Chairs – a picture sequence
15–18	Johannes Spalt Folding Chairs
19-21	Johannes Spalt Tackling the problem
22	Chairs when folded

23–42	Stools	
23–25	Tripod seat stick	1
26–28	Japanese stool	2
29–30	Shooting stool	3
31–33	Wooden stool	4
34–36	Camp stool	5
37–39	Portable stool	6
40–42	Camp stool	7
43–53	Children's chairs	
43–46	Child's stool	8
47–49	Child's folding chair	9
50–53	Child's folding chair	10
54–77	Chairs	
54–56	Garden chair	11
57–61	Boat chair	12
62–64	Boat chair	13
65–68	English folding chair	14
69–71	Fireside chair	15
72–74	Wooden chair	16
75–77	'Plia' folding chair	17
78–89	Armchairs	
78–80	Lady's rocking chair	18
81–83	Adjustable folding chair	19
84–86	Folding armchair	20
67–89	Film director's chair	21
90–92	Floor seat	22
93	Postscript	
94	List of students	
95	Literature	

Werner Blaser
Beispiel Klappstühle – eine Bildfolge

Höchstmass an Mobilität
In der vorliegenden Publikation sollen 22 ausgewählte Klappstuhltypen von den Studenten in natürlicher Grösse gezeichnet und auf Massstab $1 : 7^{1}/_{2}$ reduziert, gezeigt werden. Die Auswahl des bebilderten Teils des Vorwortes ist lediglich auf gezeichnete Darstellung konzentriert und dies mit Beispielen von den Anfängen bis in die Gegenwart. Die meisten Beispiele haben eine grosse Serienherstellung hinter sich. Der Klappstuhl wurde immer als ein mobiles und preiswertes Möbelstück angesehen. Besonders heute, wo das Einfachmöbel wieder einen höheren Stellenwert bekommt, kann die Beschäftigung mit Klappstühlen von Wichtigkeit sein. Zudem möchten manche der dargestellten Beispiele unserer Wohnforderung von mehr Flexibilität zweifellos wissenswerten Konstruktionsversuchen dienen.
Denken wir zum Beispiel an den unentbehrlichen Regenschirm. Er ist ein sinnvoller, vollkommener, funktioneller Apparat, elegant und praktisch. Wir können uns heute dieses praktikable faltbare Gerät in unserer Zivilisation nicht mehr wegdenken. Ein anderes historisches Beispiel finden wir im japanischen Haustyp: der aus sogenannten Wandschirmen mit Papierwänden, auf rechteckiggeteilten dünnen Rahmen zwischen Holzpfosten gespannt ist. Viele der Schirme sind beweglich, können zur Seite gleiten, so dass sich das Interieur je nach Wunsch verändert. Dieser ostasiatische Wesenszug der vollkommenen Mobilität war inspirierend für die westliche Wohnform des zwanzigsten Jahrhunderts. Ähnliches finden wir in den Systemen der Klappstühle, wobei Leichtigkeit, Mobilität und zugleich Originalität das Design beherrschen.

Humane Forderungen
Schon Ende der zwanziger Jahre haben die Architekten Heinz und Bodo Rasch, parallel zum Bauhaus eine Publikation „Der Stuhl", eine Darstellung aus dem Ergebnis ihrer Werkstätten herausgebracht. In dieser Veröffentlichung wird das Wesen des Klappstuhles als die organische Beschaffenheit seiner Konstruktion postuliert. „Ähnlich wie sich ein Mensch zusammenkauert, lässt sich der Stuhl zusammenlegen. Wie die Kauerstellungen des menschlichen Körpers verschieden sein können, so sind die Möglichkeiten des Zusammenklappens beim Stuhl verschieden". Obwohl vielfach die altüberlieferten Systeme des Klappstuhles beibehalten wurden, finden wir heute in Verbindung mit funktionalen Verbesserungen neue sinnvolle konstruierte und elegant fauteuilartige Typen, aufgebaut auf den Prinzipien humaner Wertmassstäbe.

Privilegierter Sitz
Der Klappstuhl ist seit alters auf das einfachste Konstruktionssystem eines x-förmigen Sägebocks zurückzuführen. Später im alten mesopotamischen Reich und in antiken Kulturen gab es schon den Klappsessel. Der Klappstuhl war auch das am allgemeinsten verbreitete Sitzmöbel des griechisch-römischen Hausrats. Der Faltstuhl im Mittelalter war Sitz hoher Würdenträger und Feldherren. Aber auch der sogenannte „Scherenstuhl", ein spätmittelalterliches Element, war aus schmalen zusammenschliessenden Latten gebaut, der beim Zusammenklappen eine enge Fläche bildet.
Aus unserem Zeitalter kennen wir den guten, alten Liegestuhl mit Segeltuch bespannt. Er ist mit wenigen Griffen zusammenzulegen oder mit mühelosen Handgriffen aufgebaut um ihn wieder platzsparend zusammenzuklappen. Heute ist die Reduktion der Masse im zusammengeklappten Zustand von grosser Wichtigkeit. Der italienische Plia-Stuhl von Piretti z. B. ist bei nur 5 cm im zusammengeklappten Zustand funktionell und raumsparend.

Werner Blaser
Example Folding Chairs – a picture sequence

Maximized mobility
In the following publication we shall show 22 types of folding chair drawn full-scale by the students and then reduced to a scale of 1 : 7½. The selection of the illustrated section of the foreword is concentrated solely on drawings and is concerned with examples from the very beginning down to the present. Most of the examples have been manufactured in large series. The folding chair has always been regarded as a mobile and inexpensive piece of furniture. Today, in particular, when simple furniture is again very much in vogue, folding chairs once more repay close attention. Moreover, at a time when we are seeking more flexibility in our living arrangements, some of the examples shown here may very well prompt worthwhile design experiments.
Take the indispensable umbrella as an example. It is a perfect piece of functional apparatus, rationally designed, elegant and convenient. It is difficult to imagine our civilization without this handy folding device. For another historical example we can turn to the Japanese type of house: the structure consists of wall screens made of paper stretched over thin rectangularly divided frames placed between wooden posts. Many of these screens are mobile and can be slid aside so that the interior can be arranged at will. This complete mobility, so characteristic a feature of East Asia, has been an inspiration to Western interior design in the 20th century. We find something similar in the systems of folding chairs where lightness, mobility and at the same time originality are dominant features of the design.

The human element
As long ago as the late twenties the architects Heinz and Bodo Rasch brought out a publication, parallel to the Bauhaus, called 'Der Stuhl' (The Chair) in which they showed the production of their workshops. In this work it was claimed that the essence of the folding chair lies in the organic character of its design. 'Just as a man can huddle himself up, so the chair can be folded together. And as the human body can be huddled up in different ways, so there are different ways of folding a chair.'
Although in many cases the old traditional system of the folding chair has been retained, we find that today functional improvements have bred new rationally conceived designs and elegant types of armchair built to the human scale.

A privileged seat
From the remotest time the folding chair has been based on the very simple design of an X-framed sawing trestle. Later there were even tip-up seats in the ancient empire of Mesopotamia and other old civilizations. The folding chair was also the commonest form of seating in the Graeco-Roman home. In the Middle Ages the folding chair furnished dignitaries and army commanders with a seat. The so-called 'scissors chair', a construction of the Late Middle Ages, was made of narrow slats interlaced so as to fold with the result that the chair took up little space when closed. In our own day we are familiar with the good old deckchair covered with canvas. It is no trouble at all to set up and can be as easily folded and put away again, taking up the minimum of space when stored. Today great importance is attached to dimensions in the folded position. The Italian Plia chair, by Piretti, is of functionally sound design and is only 5 cm thick when collapsed.

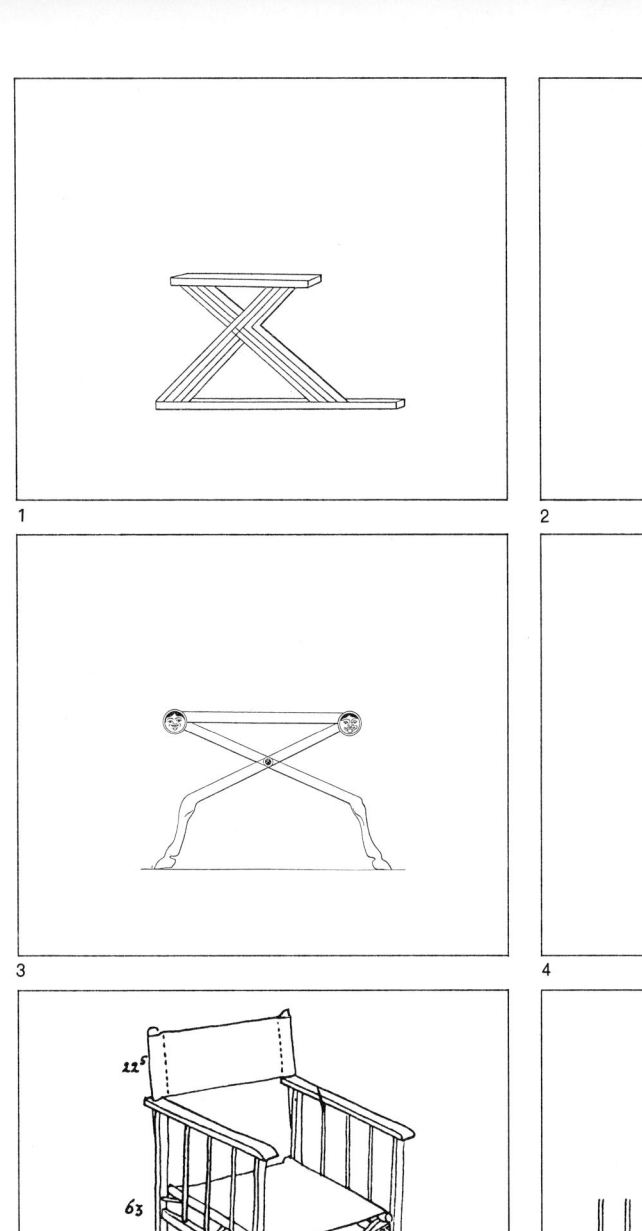
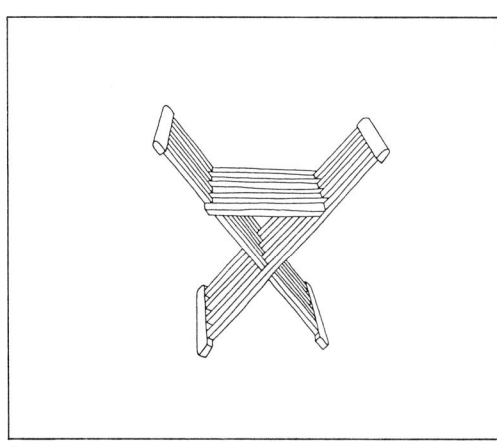
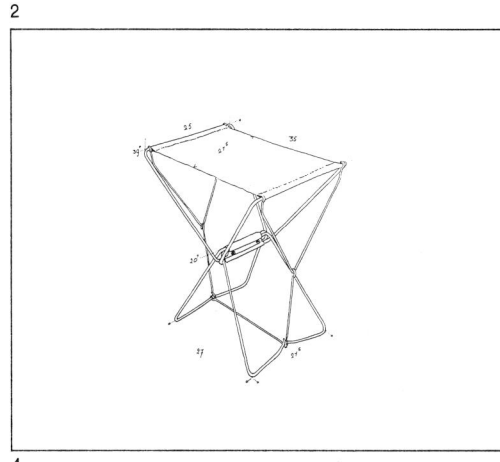
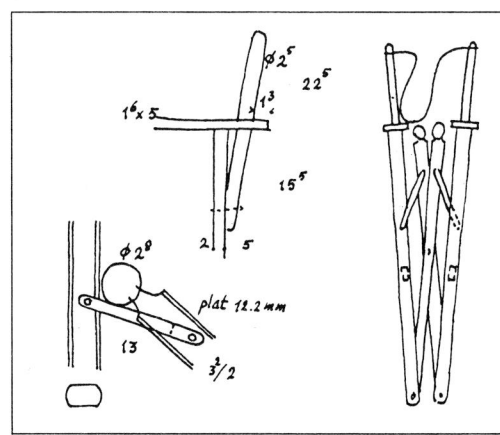

Einteilung nach der Richtung der Verkleinerungsmöglichkeit
1–6
1. Hauptgruppe: Klapprichtung normal zur Sitzvorderkante. Drehpunkt liegt unter der Sitzfläche, keine Vertikalverstrebung
1
Feldstuhl nach einem Vasenbild (griechisch-römisch)
2
Lattenstuhl (klassisches Altertum)
3
Klappstuhl nach einem Vasenbild (griechisch-römisch)
4
Mitnehmehocker gefaltet auf 25 x 18 cm
5-6
Kreuzpunkt unter der Sitzfläche mit zwei fixen Seitenteilen Regiesessel 1934

1–6
Classification according to the means employed for folding
First main group: folding direction at right angles to front edge of seat. Pivot is located under the seat, no vertical bracing
1
Camp stool from a vase decoration (Graeco-Roman)
2
Slat chair (classical antiquity)
3
Folding stool from a vase decoration (Graeco-Roman)
4
Portable stool folded to 25 x 18 cm
5–6
Intersection under the seat with two fixed side parts, film director's chair 1934

6

7–12
2. Hauptgruppe: Klapprichtung ist parallel zur Sitzvorderkante Vorderbeinstrebe durchgehend als Rückenlehne

7–9
Armlehnstuhl, Stuttgart um 1928
10
Gartenstuhl
11
Armlehnstuhl
12
Liegestuhl mit Drehpunkt unter Sitzfläche

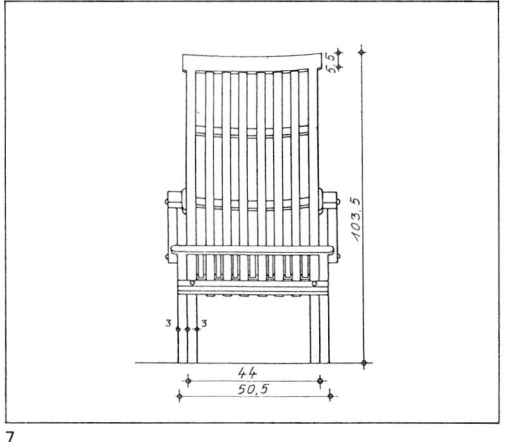
7

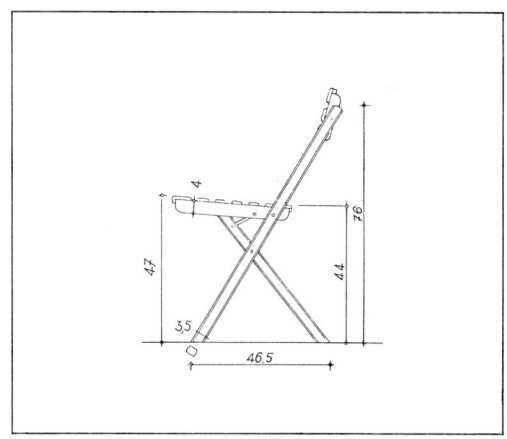
10

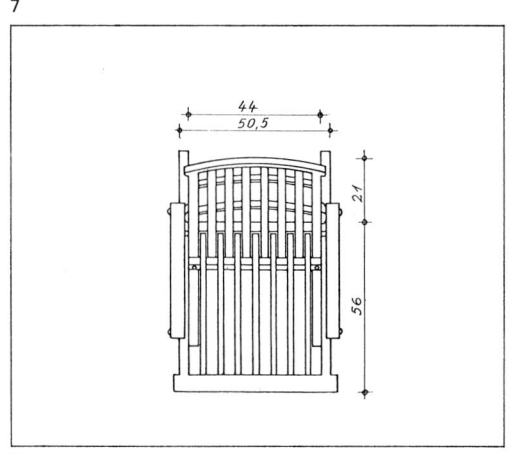
8

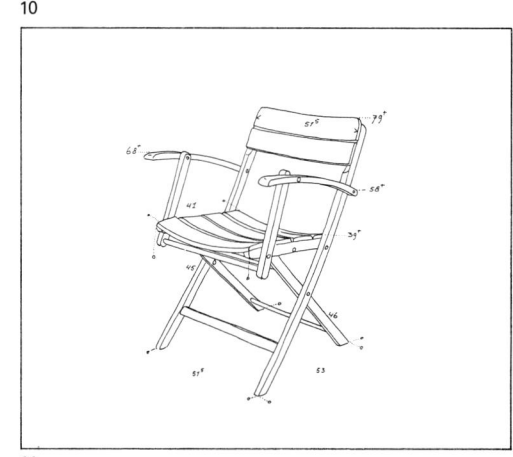
11

7–12
Second main group: folding direction parallel to the front edge of seat, front leg continuing into back rest

7–9
Armchair, Stuttgart c. 1928
10
Garden chair
11
Armchair
12
Reclining chair with pivot below seat level

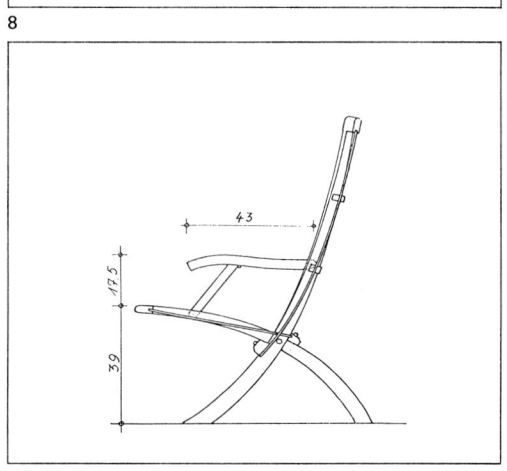
9

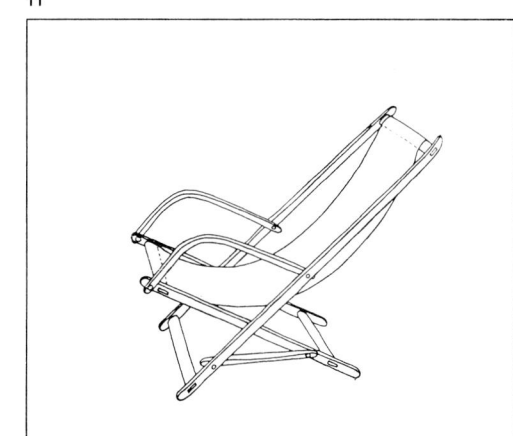
12

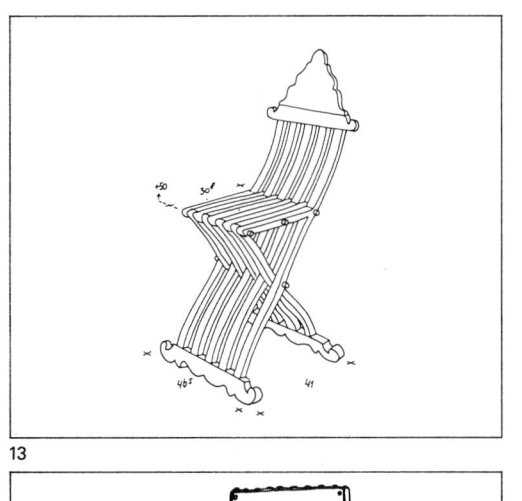

13

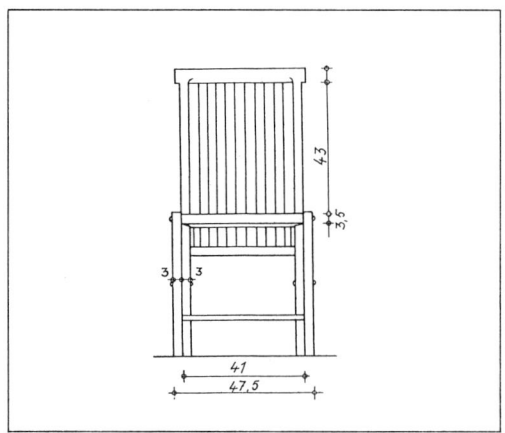

16

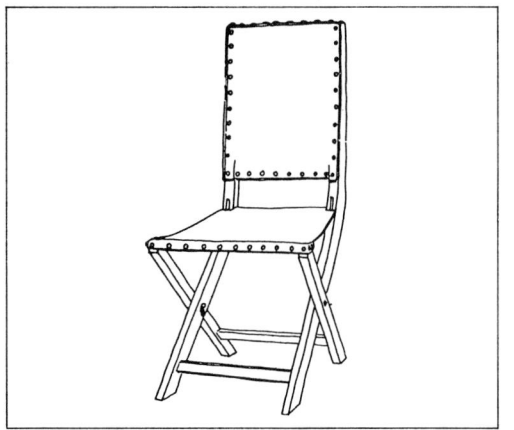

14

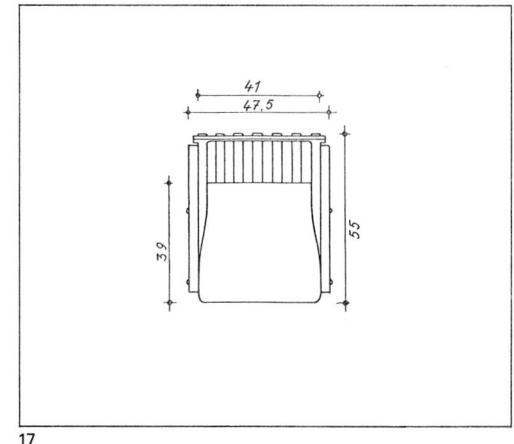

17

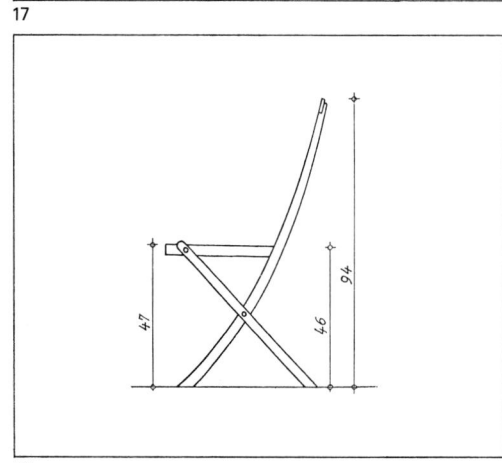

18

13–18
Drehpunkt liegt unterhalb der Sitzfläche
13
Scherenstuhl Italien 15.–16. Jh.
14
Klappstuhl Italien 16. Jh.
15
Klappstuhl
16–18
Zusammenlegbarer Holzstuhl von A. G. Schneck um 1928

13–18
Pivot below seat level
13
Savonarola chair, Italy, 15th–16th century
14
Folding chair, Italy, 16th century
15
Folding chair
16–18
Knockdown wooden chair by A. G. Schneck c. 1928

19–24
Hinterbein und Sitzfläche sind fix verbunden. Drehpunkt zwischen beiden
19–21
Gartenstühle, Stuttgart um 1935
22
Klappstuhl aus Metall
23
Stuhl mit Metallbänder
24
Gartenstuhl in Metall

19–24
Back leg and seat are rigidly connected. Pivot between the two
19–21
Garden chairs, Stuttgart c. 1935
22
Folding chair of metal
23
Chair with metal mesh
24
Garden chair of metal

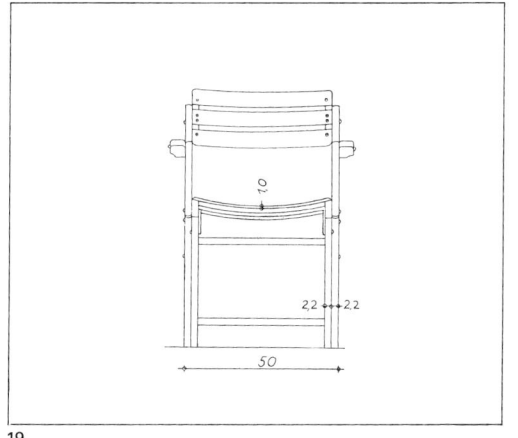
19

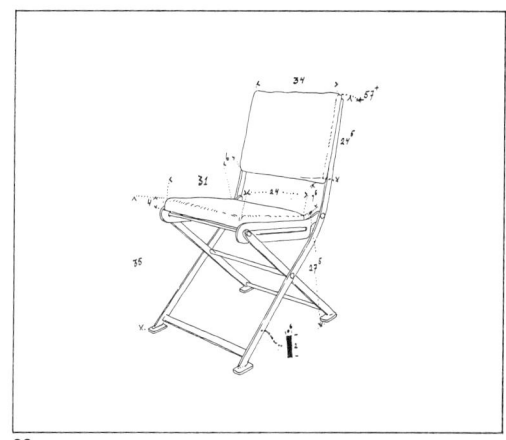
22

23

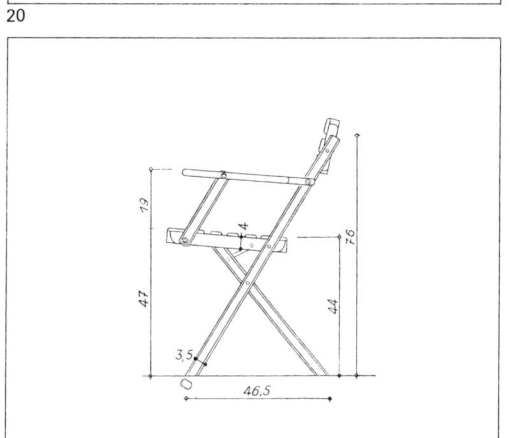
21

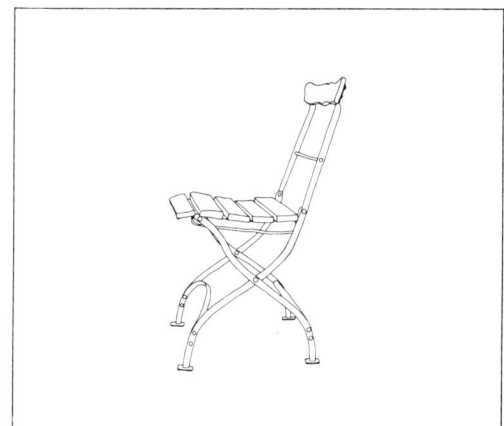
24

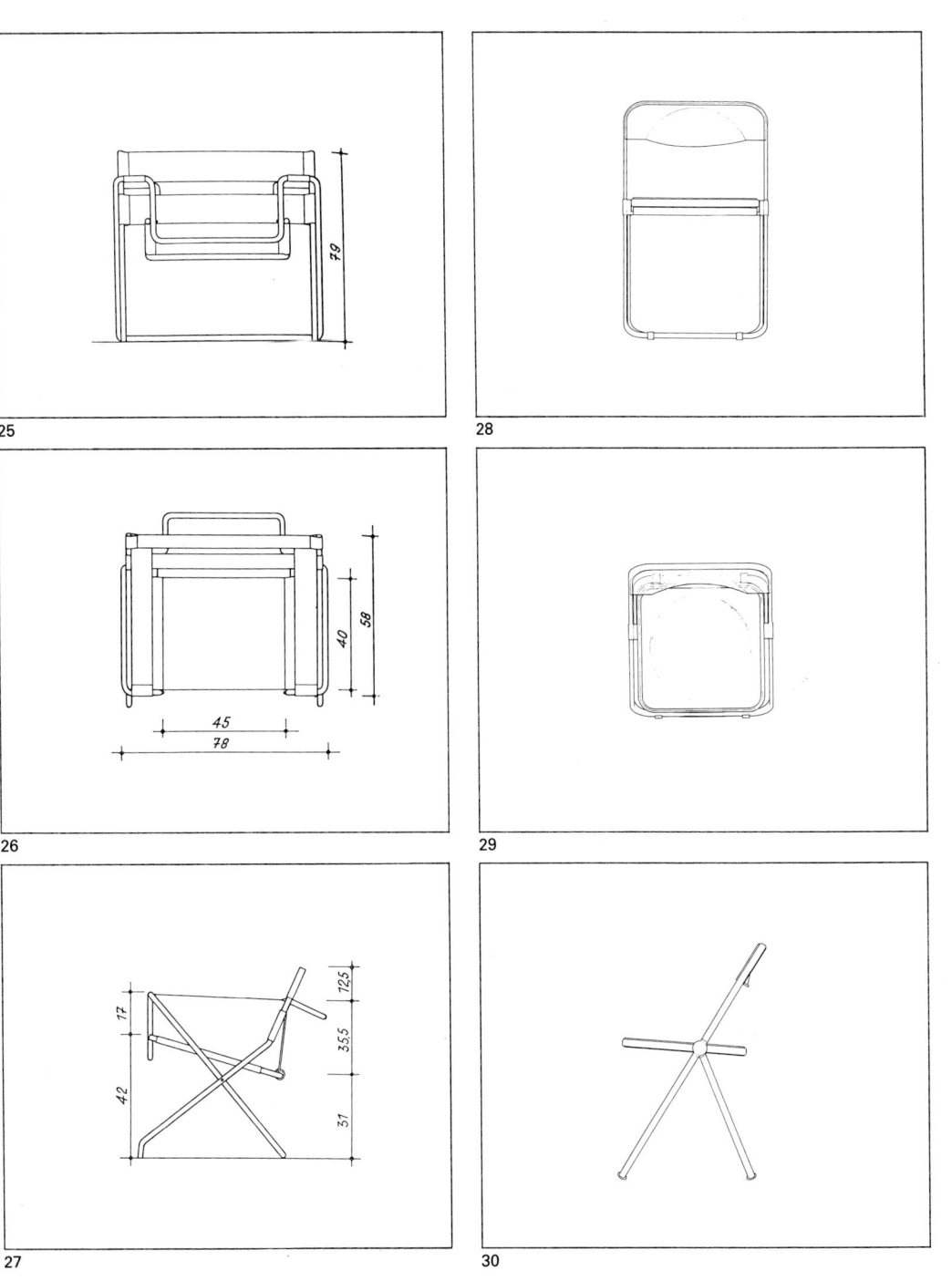

25–27
Armsessel von Marcel Breuer, Berlin um 1928 mit Drehpunkt unter der Sitzfläche
28–30
Stühle mit Drehpunkt über der Sitzfläche
Sitz- und Fussteile bewegen sich in einem Punkt
‚Plia'-Stuhl von G. Piretti Bologna 1970

25–27
Armchair by Marcel Breuer, Berlin c. 1928 with pivot below seat level
28–30
Chairs with pivot above seat level
Seat and legs hinge at one point
'Plia' chair by G. Piretti, Bologna 1970

31–36
Vorderbeinstrebe und
Rückenlehne durch Armstütze
verbunden
31
Liegestuhl mit Joncgeflecht
32
Gartenliegestuhl
33
Campingstuhl
34
Klappstuhlpatent München 1881
E. A. Naether in Zeitz
35–36
3. Hauptgruppe: Parallel und
normal zur vorderen Sitzkante
klappbar
35
Hardoy-Sessel von Bonet, Ferran,
Hardoy USA 1938
36
Segeltuchsessel mit
Aluminiumbeschlägen

31–36
Front stretcher and back rest
connected by arm support
31
Reclining chair with rattan
32
Garden lounge
33
Camping chair
34
Folding chair patent, Munich 1881,
E. A. Naether at Zeitz
35–36
Third main group: foldable
parallel and at right angles to
front edge of seat
35
Hardoy chair by Bonet, Ferran,
Hardoy USA 1938
36
Canvas chair with aluminium
fittings

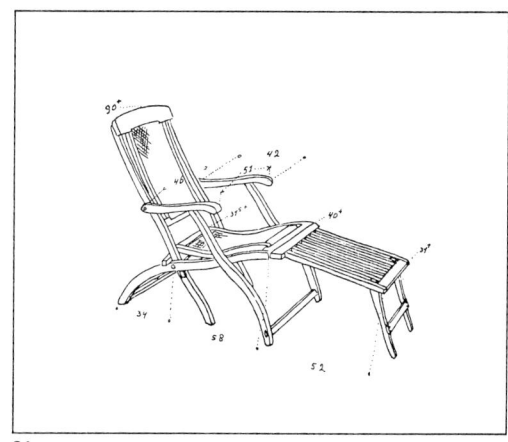
31

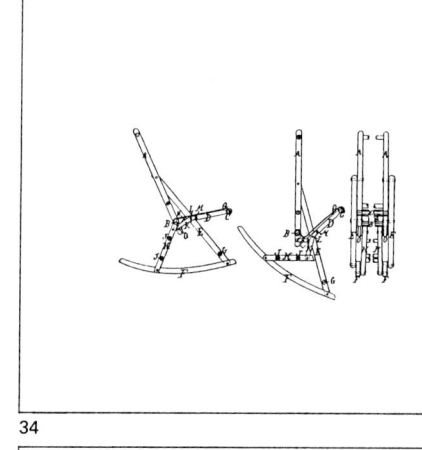
34

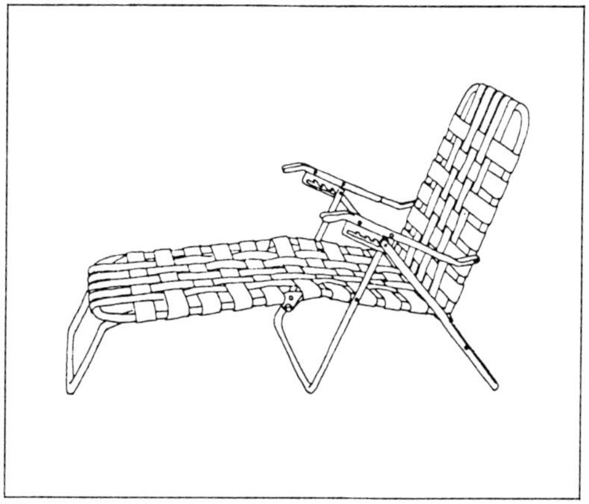

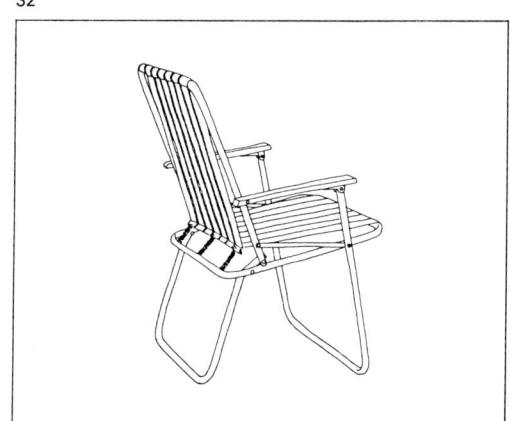
32

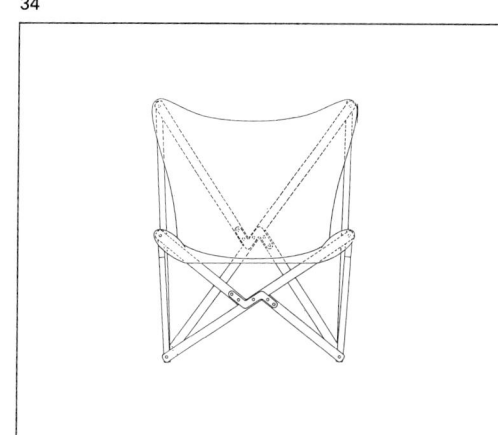
35

33

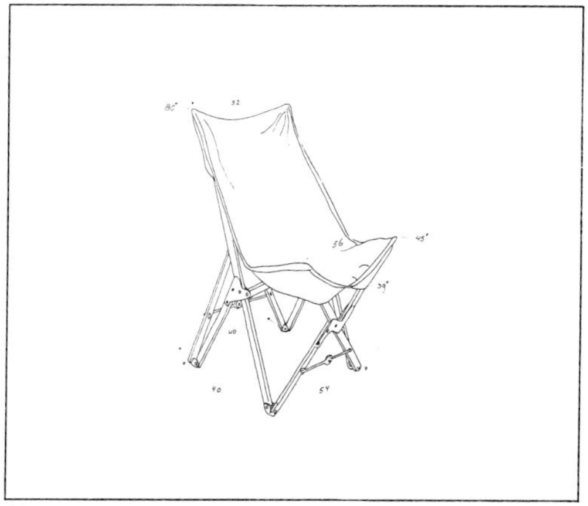

36

Johannes Spalt
Klapp- oder Faltmöbel

Unser heutiges Leben ist ohne Klapp- oder Faltmöbel in den verschiedensten Formen oder Arten kaum denkbar. Hat doch gerade die „Mobilität" in verschiedenen Ländern in Europa und Amerika so zugenommen, dass eine große Industrie für die besten, praktischsten, leichtesten Geräte sorgt und mit Auto, Schiff oder am Berg jedes Klima und Gelände ohne besondere Behinderung ertragen lässt. Überall sind Zeltstädte mit Wohnwägen oder Wohnanhängern entstanden, damit wird der Mobilität kaum eine Grenze gesetzt. Alle diese Einrichtungen sind zum Teil mit mechanischen Möbel, zumindest mit einfachen Klappmöbel ausgerüstet, da der verfügbare geringe Raum zu solchen Geräten zwingt. Das Reisen, sei es mit der Bahn, dem Autobus, Auto oder Schiff ist ebenso geprägt durch die Einrichtung mit mechanischen oder klappbaren Möbel. Der Liegesitz im Auto, das Couchette in der Bahn, sind die bekanntesten Beispiele. Vorläufer dieser Lösungen waren die schönen Klappsitzkonstruktionen in den Pferdekutschen. Auch die englischen Taxis alter Bauart haben heute noch Klappsessel. Wie aber würden unsere Theater- und Kinosäle ohne die speziellen Klappsessel auskommen, die wir als selbstverständliche Einrichtung hinnehmen; trotzdem haben alle diese Möbel eine eigene Entwicklung und Geschichte, die mit anderen technischen Errungenschaften zusammenhängt. Wir könnten heute kaum den alten klassischen Liegestuhl (Deckchair) entbehren, der eine ungeheure Verbreitung erfahren hat und für geringen Preis überall zu haben ist. Man könnte ein wenig übertrieben sagen, wie leben in einer eigenen Klapp- oder Faltstuhlkultur, so sehr hat sich das Vorbild der Bewegung unserer eigenen Extremitäten auf das Möbel ausgewirkt.

a Die Beschäftigung mit dem Problem des Sessels führt zwangsläufig zum Beginn des Sitzens auf einem dazu hergestellten Gerät. Die Sitzgewohnheiten haben sich aber in den Jahrhunderten sehr geändert, so kultivierten viele Völker das Sitzen auf dem Boden, das Hocken oder Knien mit Sitz auf den Fersen, wie Japaner, Perser, Türken und viele Völker.

4 Die Hierarchie musste in früheren Zeiten herausgehoben werden. Die gekreuzten Beine symbolisiert der Falthocker. Dieser musste den seinerzeitigen Notwendigkeiten entsprechend transportabel und leicht sein. Falthocker und Faltstuhl zählen daher zu den ältesten Sitzmöbel. Eine grosse Zahl an Beispielen aus dieser Frühzeit hat sich erhalten,
5 wir kennen ägyptische, griechische, römische, asiatische und romanische x-förmige Hocker. Das aus zwei sich kreuzenden Rahmen bestehende Gestell wurde durch ein Leder- oder Stoffteil verbunden und durch einen Dorn an der Kreuzungsstelle beweglich gehalten. Auf dem Leder- oder Stoffteil hat man meist einen Polster gelegt, manchmal aber auch den Sitzteil tiefer angesetzt damit kleine Seitenlehnen entstanden. Die Enden der Gestelle waren mit Symbolen wie Entenschnabel bei den Ägyptern, Löwen oder Drachenköpfen bei den Römern geschmückt. Als Material wurde Holz, Bronze oder Eisen verwendet. An diesem Typ des Klapphockers hat sich im grundsätzlichen bis heute nichts geändert. Er ist ein Gebrauchsgegenstand. Auch statisch ist der Falthocker durch die vom Gelenkpunkt aus verkürzten Beinschenkeln günstig und erlaubt eine schlanke Dimensionierung des Gestells. Ausserdem ist der Hocker sehr standfest, da sich die Sitz-
1 fläche bei Belastung spannt. Im alten Ägypten wird der Falthocker in einfacher Form ab der
2 18. Dynastie Allgemeingut. Die griechischen Falthocker zeigen zum Unterschied zu den ägyptischen einen mehr dynamischen Ausdruck durch die sprunghaft wirkenden Löwenhinterbeine oder nach aussen gewendeten Ziegenbeine. Als Material wurde Zedernholz oder Esche verwendet, das durch seine Elasti-
3 zität besonders geeignet war. Römische Falthocker sind hergestellt aus Bronze oder Eisen, wobei die Enden der Beine nicht mehr verbunden sind. In der Mitte sind diese mit Bolzen beweglich gehalten. Dieser Punkt ist durch eine Scheibe markiert. Rippenfaltstühle oder Scherenfaltstühle und -hocker sind bereits im 4. Jh. v. Chr. in Griechenland verwendet worden. Durch die Möglichkeit der gelenkigen Lagerung der Sitzleisten auf einer Seite sind sie zusammenklappbar. Durch die Vielzahl der Rippen wurde eine grosse Stand-

festigkeit erreicht, ohne dass besonderes handwerkliches Werkzeug benützt werden mußte.

8 Klappstühle aus kammartigen ineinandergreifenden Holzleisten sind in der Frührenaissance sowohl in Deutschland wie in Italien zu finden. Sie werden auch Scheren-
7 stühle genannt. Hat sich die Richtung des x-förmigen Gestells geändert, so konnte durch Verlängerung des einen Fussteils eine Lehne erreicht werden. Diese Scherenstühle aus Leisten wurden später nach dem Sitzprofil geformt und ermöglichten ein bequemeres
6 Sitzen. Der venezianische Faltstuhl der Frührenaissance bestand aus teilweise geschnitzten brettartigen Teilen mit Lehne und Armstützen, die durch grosse Holznägel verbun-
9 den waren. Die chinesischen Faltstühle des 18. Jhdts. hatten eine bugholzartige Rückenlehne sowie Eisenbeschläge und eine Fußstütze. Ein vollkommenes Zusammenlegen war daher unmöglich.

Im 19. Jhdt., in der Euphorie der industriellen und mechanischen Möglichkeiten sind besonders in Amerika aber auch in Europa Entwürfe für mechanische Möbel entstanden, die weit über die heute existierenden in ihrer Vielfalt hinausgehen. Das System der „Tischlein deck' dich"-Zauberei hat sich teilweise
10 erfüllt. Gartensessel aus Holz und Eisen, und
11 Liegestühle vorerst als Deckchair wurden in grosser Zahl produziert. Ein neuer gedanklicher Prozess findet für den zusammenlegbaren Sessel und Hocker in hunderten Patenten in der ganzen Welt seinen Ausdruck. Es gibt seit dieser Zeit Sitzklapper, Beinklapper, Sitz- und Beinklapper, Armlehnklapper, klappbare Schaukelsessel, Schiffsessel, Kindersessel, Jagdsessel, Regiesessel, Hocker, Krankenstühle, Campingsessel, Spazierstocksessel, Kaminsessel – Modelle für jede Gewohnheit, jeden Bedarf und jede Notwendigkeit.

12 Nun, 100 Jahre später, besinnt man sich, mit neuen technischen, antriebstechnischen Möglichkeiten ausgerüstet, des damaligen Impulses. Höhepunkt der mechanischen Möbel sind heute der Zahnarzt- oder der Operationsstuhl. Er kann sich fast jeder Form menschlicher Haltung anpassen.
Für den technischen Vorgang waren einmal die Hebelgesetze, aber auch das Verwenden von Scharnieren, Achsverschraubungen, Vernietungen, Federn und die neuen Materialien wie leichtes Stahlrohr, Aluminium, Bugholz, Glasfiber, Kunststoff massgeblich. Mit den neuen Klappmöbeln war auch ein viel grösserer Grad der Anpassung an die neuen Erfordernisse möglich. Von dieser ist die Raumersparnis und das geringe Gewicht sicher das entscheidende Moment. Natürlich unterscheiden wir, was und wie und welcher Teil des Möbels bewegt wird, um die leichteste Bedienbarkeit zu erreichen. Die Vielseitigkeit
14 der Verwendung spielt eine grosse Rolle, denn nur einfache Möbel können dieses Kriterium erfüllen, wie z. B. der Hocker. So hat sich der Klappsessel und der Klapptisch, für die Gartenmöbel geeignet, besonders wegen der guten Stapelbarkeit, einem Kriterium, das auch bei Kindermöbel, bei Schiff- und Strandsessel von Bedeutung ist, durchgesetzt.
Die verschiedenen Arten von Mechanik bei bekannten Klappstühlen sollen in einigen Beispielen beschrieben werden. So wird bei Gartenklappstühlen der feste Sitz hochgeklappt; eine hebelartige Eisenstange verbindet jeweils die Fussteile und schiebt dadurch den Sitzteil hoch.
Ein anderes Prinzip lässt den Sitz durch einen Hebel, der mit den Vorderfüssen verbunden ist und bis zur halben Sitzfläche zurückreicht, hochklappen. Durch seitliche Nuten an der Sitzfläche oder einer Metallführung, die ähnliche Wirkung hat, ist der Sitz einklapp- oder hochklappbar.
Durch frontale Scheren, die mit zwei Rahmenteilen verbunden sind, wird ein anderes Prinzip vielfach verwendet. Erlaubt es doch Armteile und Lehnenbespannung günstig anzubringen.
Wird der Sitz an den durchgehenden Vorderfüssen, die auch die Rückenlehne bilden, schwenkbar gelagert und in den ausgespreizten Hinterfüssen in Nuten geführt, so ergibt sich ein brauchbares, stapelbares Möbel.
In einer neueren Konstruktion treffen Vorderfüsse und strebenartige Hinterfüsse an einem Punkt, der das Gelenk für den drehbaren Sitz aufnimmt, zusammen (Plia).
Dreibein- und Vierbeinhocker verdrehen um einen zentralen Bolzen die Beinelemente, die

Sperre erfolgt durch die Sitzbespannung oder das Sitzteil.
Im Liegestuhlprinzip werden zwei Rahmenteile so weit geschwenkt, wie die Stützstrebe auf dem zahnartigen Teil des Hinterrahmens erlaubt. Armstützen sind ebenso mitklappbar.

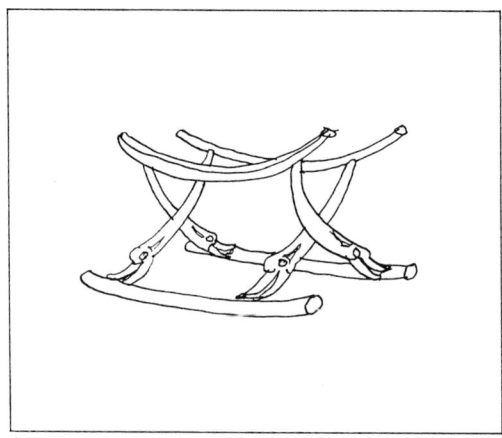

1

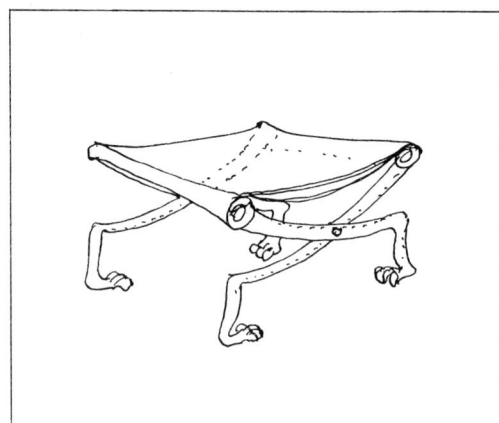

2

a

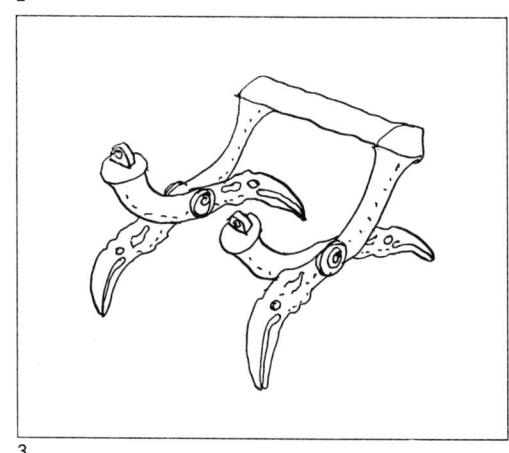

3

1
Ägyptischer Hocker
mit Entenschnäbeln, 1300 v. Chr.
2
Griechischer Falthocker,
ca. 500 v. Chr.
3
Falthocker aus Pompei,
1. Jh. n. Chr.

1
Egyptian stool with ducks' bills, 1300 BC
2
Greek folding stool, c. 500 BC
3
Folding stool from Pompeii, 1st century AD

Johannes Spalt
Folding Chairs

Life today would be almost inconceivable without folding or knockdown furniture in all its forms and variety. 'Mobility' in various countries in Europe and America has increased to the point where there is a special industry producing the best, most convenient and lightest pieces of furniture which, whether carried in cars or ships, or set up in the mountains, can cope with every climate and terrain without difficulty. Everywhere camping sites with tents, caravans and trailers have sprung up and there seem to be no limits set to mobility. All these are equipped partly with mechanical furniture and at least with simple folding units, since the small amount of space available makes such adjuncts necessary. Travel, whether by rail, bus, car or ship, is also characterized by the use of mechanical or folding furniture. The reclining seat in the car and the couchette in the train are the best known examples. The predecessors of these devices were the beautifully wrought let-down seats in the old mailcoaches. Even English taxis of the old style still have tip-up seats. And how would our theatres and cinemas manage without the special tip-up seats which we take for granted as standard equipment. Even so, all these types of furniture have their own development and history which are linked together with other technical achievements. Today we could scarcely manage without the old classic deck chair which has become so enormously widespread and costs very little. It would hardly be an exaggeration to say that we are living in a folding-chair culture — such has been the influence of our mobility on the furniture we use.

a Concern with the problem of the chair necessarily leads back to the early days when man first began to sit on a support made specially for the purpose. Sitting habits, however, have changed greatly over the centuries; many peoples made a practice of sitting on the ground or squatting on their heels, for example the Japanese, Persians, Turks and many other nations.

In olden days the hierarchy had to be given 4 prominence. The folding stool symbolizes crossed legs. To meet contemporary conditions the stool had to be light and readily transportable. Folding stools and chairs are therefore some of the oldest forms of seating. A large number of examples from earlier times have been preserved; we know
5 Egyptian, Greek, Roman, Asiatic and Romanesque X-shaped stools. The two frames forming the stool were secured by a pin where they intersected so that they could be moved; fabric or leather was used to join the tops of the frames, but sometimes the seat was placed lower so that small armrests were formed at the sides. The ends of the frames were decorated with symbols such as ducks' beaks in the case of the Egyptians, and the heads of lions or dragons among the Romans. The materials used were wood, bronze or iron. There has been no essential change in this type of folding stool down to the present day. It is an object of everyday use. Since the legs are shortened by the joint, the folding stool is statically sound and the members of the frame can be kept slim. Moreover, the stool is very stable because the seat comes under tension when weight is placed upon it.
1 In ancient Egypt the folding stool in its simple form was widely known from the XVIIIth dynasty onward. Compared with the Egyptian
2 examples, the Greek folding stool creates a more dynamic impression because of the hindlegs of lions which seem to be about to spring, and the outward splaying of goats' legs. The material used was cedar or ash, which was particularly suitable because of its
3 resilience. In Roman times iron and bronze were already being used for making stools, but in these examples the ends of the legs are not joined together. They are secured in the middle with bolts to form a movable joint. This point is marked with a disk. Scissors chairs and stools with interlaced slats were known in Greece as early as the 4th century BC. The slats of the seat are hinged on one side and thus enable it to be folded up. The number of slats gives the whole extra stability without requiring the use of special tools.
8 Folding chairs consisting of wooden slats interlaced like combs were to be found in both Germany and Italy in the Early Renais-

sance. They are also known as scissors chairs. When the disposition of the X-shaped frame was changed, one of the legs could be continued into a back rest. These scissors chairs of slats were later shaped to the sitting profile and this made for greater comfort. The Venetian folding chair of the Early Renaissance consisted of sometimes carved board-like elements with a back and arms which were held together by wooden dowels. Chinese folding chairs of the 18th century had a bentwood-like back with iron fittings and a footrest. They could therefore not be completely folded.

In the 19th century, during the euphoria created by the vast expansion of industrial and mechanical facilities, designs were produced in America and also in Europe for mechanical furniture which, in the range of its diversity, greatly exceeded anything existing today. The fulfilment of wishes by a kind of Cockaigne-like magic was partly achieved. Garden chairs of wood and iron and reclining chairs, at first of the deck chair type, were produced in large numbers. Novel ideas for folding chairs and stools were reflected in hundreds of patents all over the world. Since that time there have been folding seats, folding legs, folding seats and legs, folding armrests, collapsible rocking chairs, boat chairs, children's chairs, shooting stools, film director's chairs, stools, invalid chairs, camping chairs, shooting sticks, fireside chairs, in fact chairs for every habit, need and necessity.

Today, 100 years later, with our new technical resources and mechanical drives, we are reviving some of these ideas. Today the acme of mechanical furniture is to be found in the dentist's and operation chair which can adapt itself to almost any posture of the human body.

Technical requirements were formerly met by systems of levers and also the use of hinges, screwed pivots, rivets and springs while today light tubular steel, aluminium, bentwood, fibreglass and plastics are used. The new folding chairs allow a much higher degree of adjustment to novel conditions, prompted mainly by a desire to save space and to lighten weight. Needless to say, we make a distinction as to what, how and which part of the furniture is moved so as to make for the greatest possible ease of operation. Versatility is also of the first importance, for only simple furniture, such as the stool, can satisfy this requirement. Thus the folding chair and folding table, suitable for garden furniture, have proved popular because of their stackability, a feature which is also of importance in children's furniture, and in beach and boat chairs.

A few examples will be quoted to illustrate the different types of mechanism used in well-known folding chairs. In garden folding chairs, for instance, the fixed seat is tipped up; a lever-like iron rod connects the legs and serves to raise the seat. Another principle allows the seat to be raised by a lever which connects the front legs and extends back up to half the height of the seat. Grooves in the sides of the seat or a metal guide, which has a similar action, enable the seat to be tipped up or down.

Another principle which is commonly used involves frontal scissors which are connected with two frames. With this system arms can be readily fitted or the back can be covered. If the seat is pivoted on front legs which continue into the back and slides along grooves in the splayed back legs, the chair can be readily stacked.

In a more modern design the front legs and strut-like back legs meet at a point which embodies the articulation for the pivoting seat (Plia).

Three-legged and four-legged stools impart a turning motion to the legs round a central pivot, a locking action being provided by the covering or the seat. In the deck chair principle two pivoted frames open as far as the stretcher engaging in the tooth-like back frame allows. Armrests can also be folded.

4
Eiserner Faltstuhl aus England, 6. Jh.
5
Bischofssitz (faldistorium), Frankreich, 14. Jh., aus geschmiedetem Eisen
6
Venezianischer Faltstuhl, Frührenaissance
7
Scherenstuhl, Frührenaissance Deutschland
8
Scherenstuhl, Schweiz, um 1500
9
Chinesischer Faltstuhl, 18. Jh.

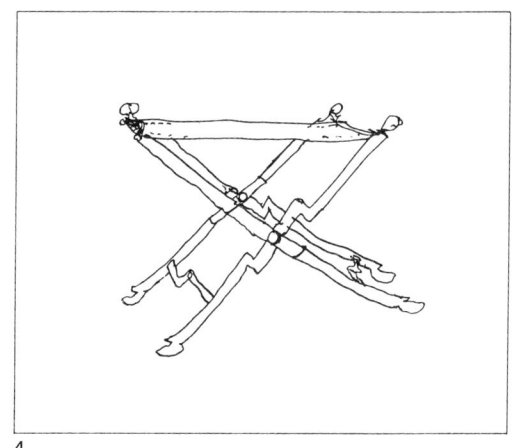

4

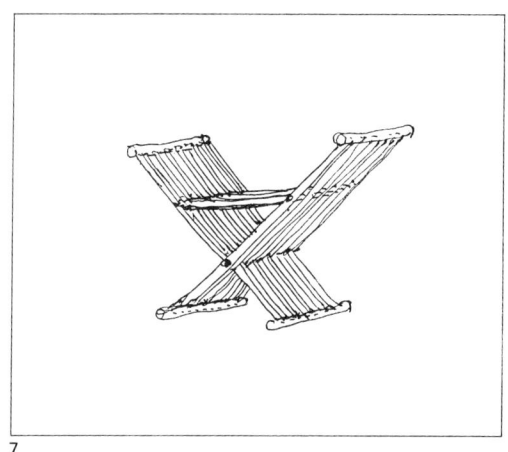

7

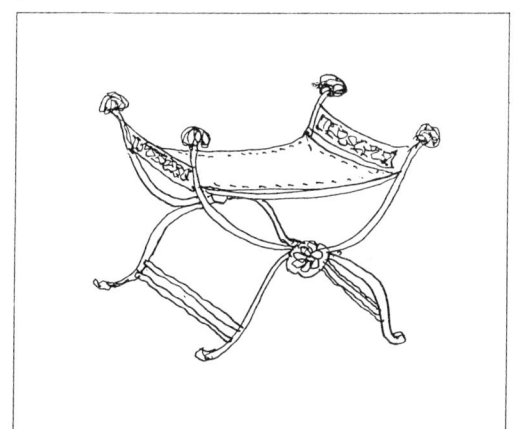

5

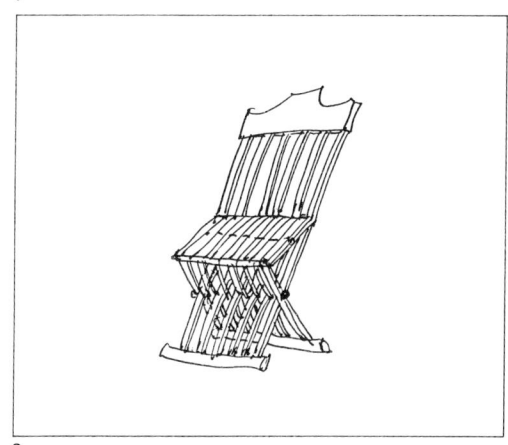

8

4
Iron folding stool from England, 6th century
5
Bishop's seat (faldistorium), France, 14th century, of wrought iron
6
Venetian folding stool, Early Renaissance
7
Scissors chair, Early Renaissance, Germany
8
Scissors chair, Switzerland, c. 1500
9
Chinese folding chair, 18th century

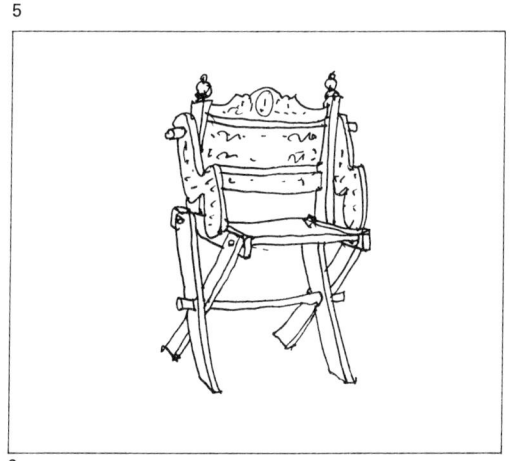

6

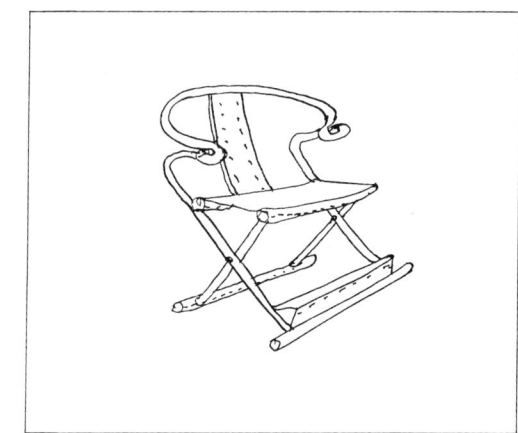

9

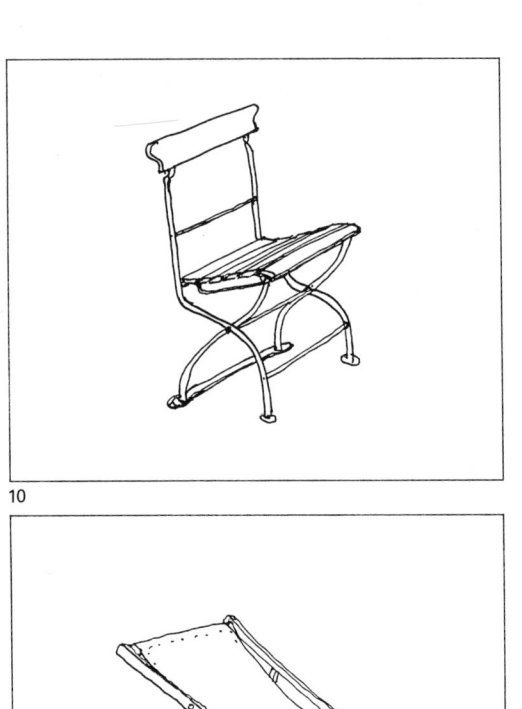

10

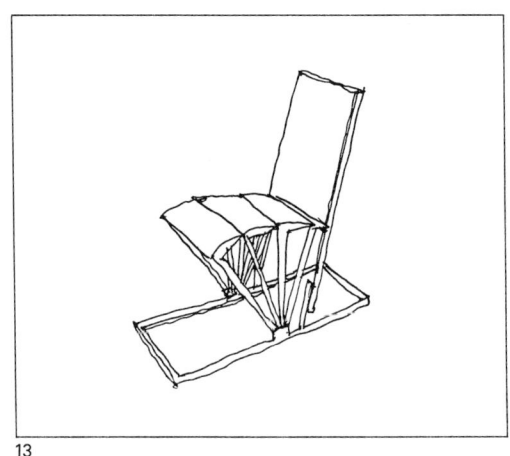

13

10
Gartensessel aus Eisen,
Mitte 19. Jh.
11
Liegestuhl (Deck chair),
2. Hälfte 19. Jh.
12
Fauteuil Stahlrohr vernickelt,
zusammenklappbar,
Marcel Breuer 1926
13
Stuhl aus Bandeisen lackiert,
Pierre Chareau 1930
14
Dänischer Klappstuhl,
Hans Wegener 1949
15
Dänischer Schiffsfaltstuhl,
Torsten Johansson 1953

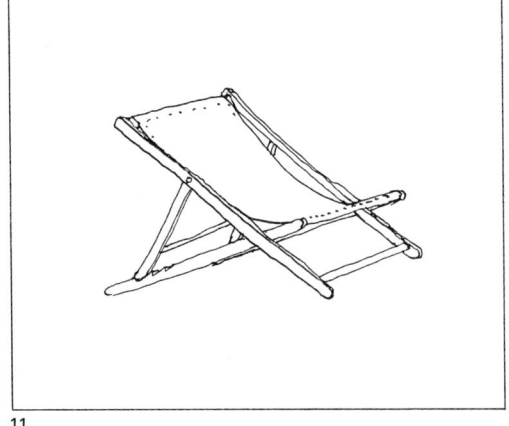

11

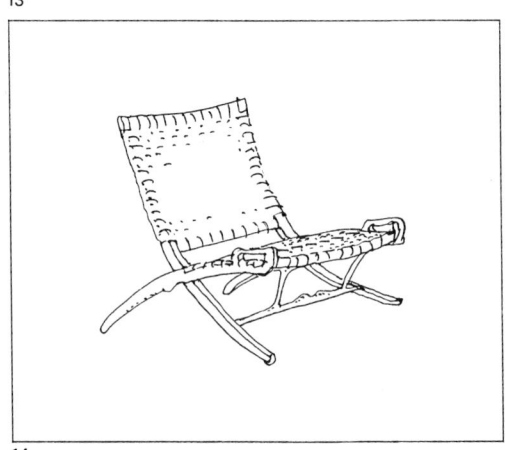

14

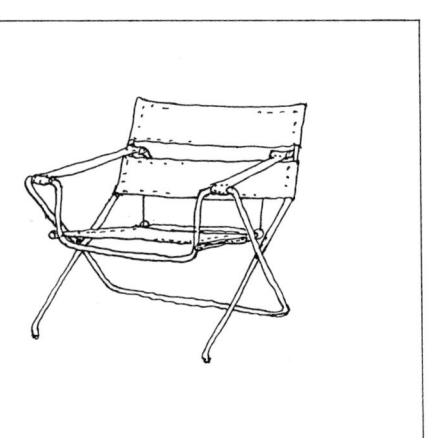

12

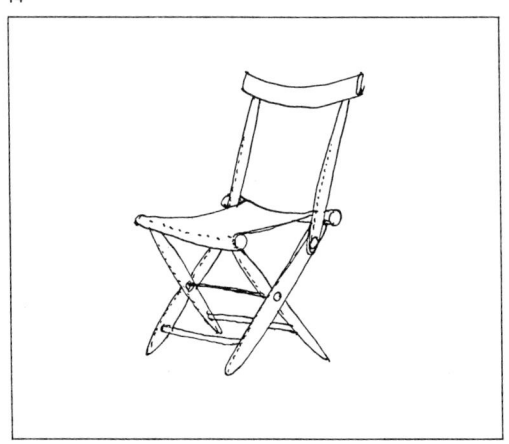

15

10
Garden chair of iron, mid-19th century
11
Deck chair, 2nd half of 19th century
12
Nickel-plated tubular steel chair, collapsible, Marcel Breuer 1926
13
Chair of japanned strip steel, Pierre Chareau 1930
14
Danish folding chair, Hans Wegener 1949
15
Danish folding chair for shipboard use, Torsten Johansson 1953

Johannes Spalt
Aufgabenstellung

Es war eine aufregende Sache, als wir in der Meisterklasse mit wenigen vorhandenen Originalsessel das Problem zu studieren begannen. Es haben sich aber in kürzester Zeit durch das Interesse der Studenten und Lehrkräfte alle wesentlichen Typen angesammelt. Der Gedanke, durch Aufmass und Studium der Systeme zu neuen Produktentwürfen in relativ kurzer Zeit zu kommen, hat sich nur teilweise erfüllt. Denn die meisten dieser vorhandenen Sessel wurden von Aussenseitern entwickelt. Es ist eine weit verbreitete Meinung, neue Sessel seien schnell und einfach zu entwerfen. Das ist ein Irrtum.
Sessel und besonders Klappsessel gehören zu den schwierigsten Entwurfsprogrammen. Klappsessel werden durch ihren notwendigen Klappmechanismus ausserdem teilweise in ihrer Sitzqualität beeinträchtigt, was möglichst vermieden werden muß. So verblieb von diesem Programm eine Anzahl Massaufnahmen und Fotos vorhandener Stühle, an denen die wesentlichen Systeme erkannt und abgewandelt werden konnten. Diese Massaufnahmen herauszugeben und für einen grossen Kreis wirksam zu machen, war bereits am Beginn der Arbeit unsere Absicht. Natürlich stellt dieser Ausschnitt nur den Anfang einer Sammlung dar, die die Vielfalt und den gedanklichen Reichtum einer Entwicklung aufzeigen soll, ist doch der Stuhl das uns am nächsten kommende Möbel – ein negatives Abbild, eine Stütze, eine Entlastung, ein Gerät. „Klappstühle" vergrössern unseren Raum, wir können sie wegstellen und gewinnen Platz.

Johannes Spalt
Tackling the problem

It was very exciting for us in the master class when we began to study the problem with only a few original chairs available. However, examples of all the most important types were quickly assembled through the interest of the students and teaching staff. The idea that measurements and study of the systems would soon lead to new product designs has only been fulfilled in part; for most of the chairs available were designed by outsiders. It is a common misconception that new chairs are easy and quick to design. Nothing can be farther from the truth. Chairs and folding chairs in particular are among the most difficult design programmes. Sometimes the indispensable folding mechanism detracts from the comfort of folding chairs, and this must be avoided as far as possible. Thus there remained from this programme a number of dimensioned drawings and photographs of existing chairs in which the most important systems could be identified and developed into variant forms. It was the aim at the very beginning of our work to publish these dimensioned drawings and make them available to a wider public. Needless to say this selection is only the beginning of a collection which will show the diversity of the chair and the wealth of ideas embodied in it, for it is, after all, the piece of furniture with which we are most intimately connected – a negative image, a support, a relief, an appliance. 'Folding chairs' enlarge the space we live in; they can be put away and leave us more room.

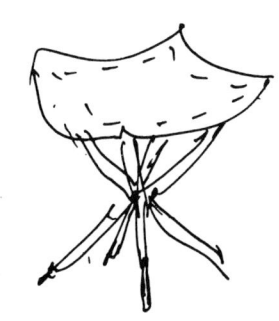

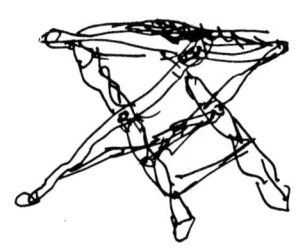
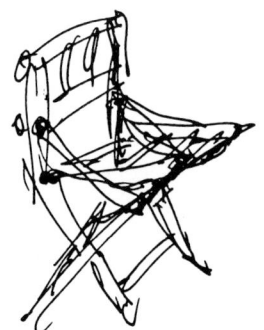

1
Damenstockhocker
Thonet Österreich (Stocksessel
Nr. 3/6823) Sitz am Stock auf-
klappbar und mit einer Messing-
hülse arretierbar, Bugholz natur,
Sitz mit Rohrgeflecht
Sitzdurchmesser   23 cm
Gesamthöhe   53 cm
Spazierstock   86 cm

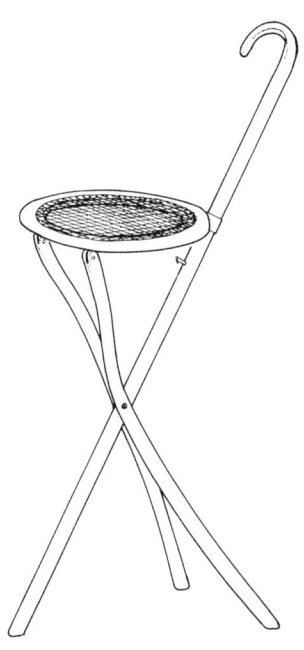

1
Tripod seat stick
Thonet, Austria (seat stick
No. 3/6823) folding seat on stick,
brass sleeve locking device,
bentwood natural finish, cane
seat
Diam. of seat   23 cm
Overall height   53 cm
Walking stick   86 cm

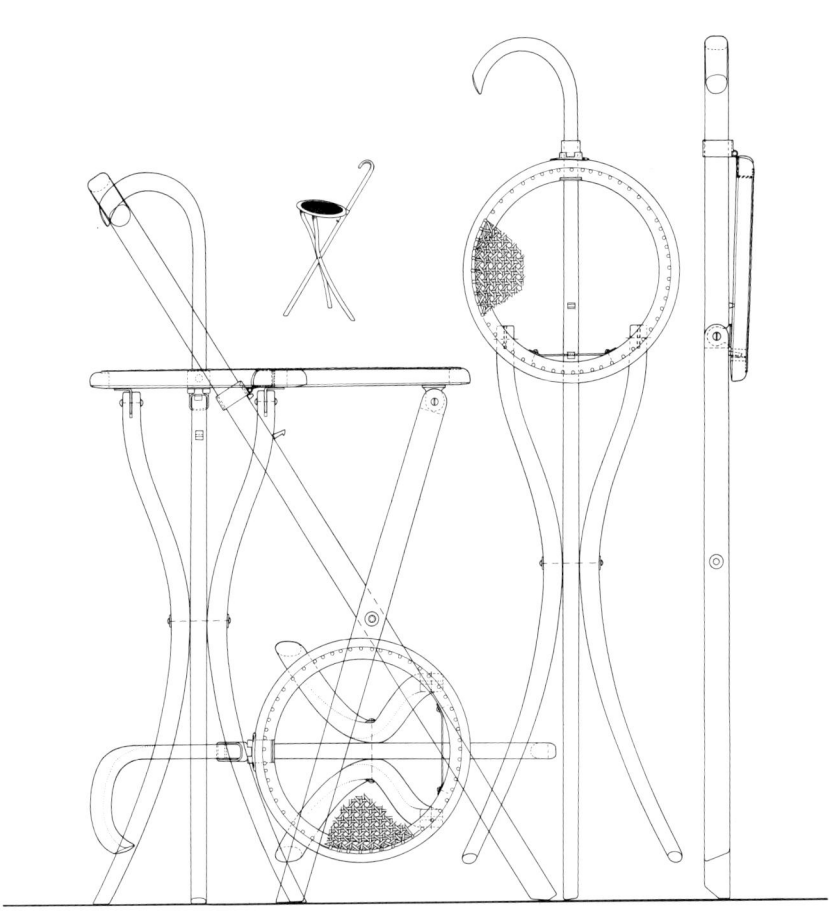

Alle Zeichnungen im Massstab 1:7½
25 All drawings on a scale of 1:7½

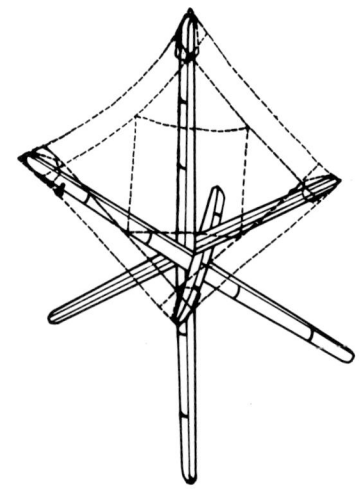

2
Japanischer Hocker
aus einem vierteiligen aufspreizbaren Stab und einem unabhängig mit Gurten verstärkten Plachenstoffsitz zum Aufschlupfen. Stab aus Buche mit kreuzförmigem Gelenk aus Stahl in Bambusimitation

Sitzhöhe	35 cm
Sitzfläche	40 x 40 cm
Stabdurchmesser	2,5 x 3,5 cm
Geschlossen	52 cm

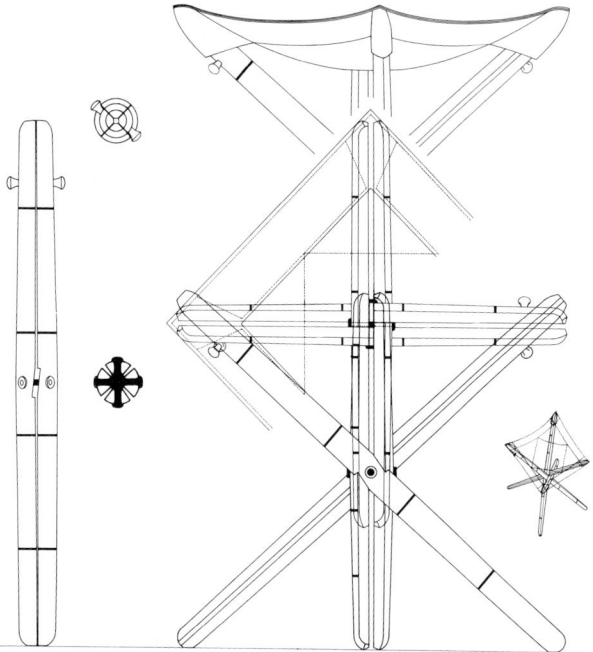

2
Japanese stool
The 'stick' comprises four parts which are opened to take a separate web-reinforced canvas seat. Stick of beech with cruciform joint of steel in imitation of bamboo

Height of seat	35 cm
Area of seat	40 x 40 cm
Diam. of stick	2.5 x 3.5 cm
Closed	52 cm

3
Jagdstockerl
Thonet Österreich (Stöcke
Nr. 2/6802) um 1900
gedrechselte Buchenfüße mit
Ledersitz. Der dreieckige Sitz ist
mit Ledergurten verstärkt und an
den Stöcken mit Messing-
schrauben verbunden
Sitzhöhe   46 cm
Kantenlänge   26 cm
Gefalten   56 cm

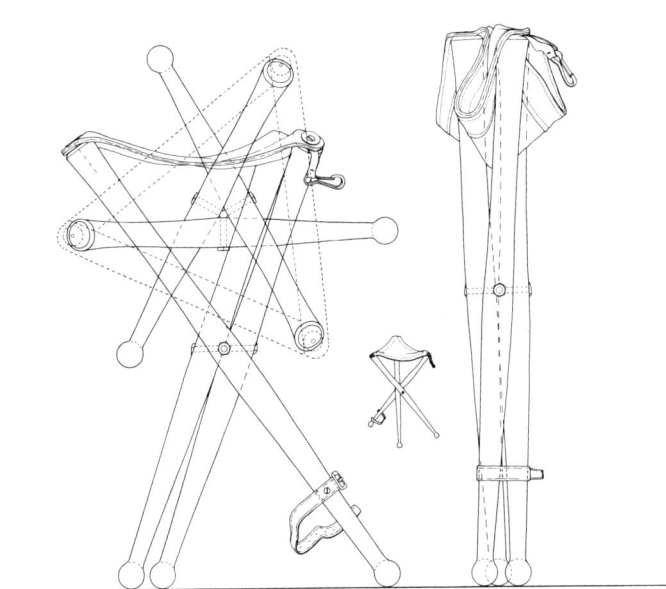

3
Shooting stool
Thonet Austria (No. 2/6802)
c. 1900 turned beech legs with leather seat. The three-cornered seat is reinforced with leather straps and secured to the sticks with brass screws

Height of seat	46 cm
Length along edge	26 cm
Folded	56 cm

4
Holzstockerl
aus Buche mit geteiltem Sitz der
hochklappbar ist
Sitzhöhe   38 cm
Sitzfläche   29 x 31 cm
                    (tief)
zusammengelegt   59 cm

4
Wooden stool
of beech with a seat which folds
in two and collapses
Height of seat 38 cm
Area of seat 29 x 31 cm
Folded 59 cm

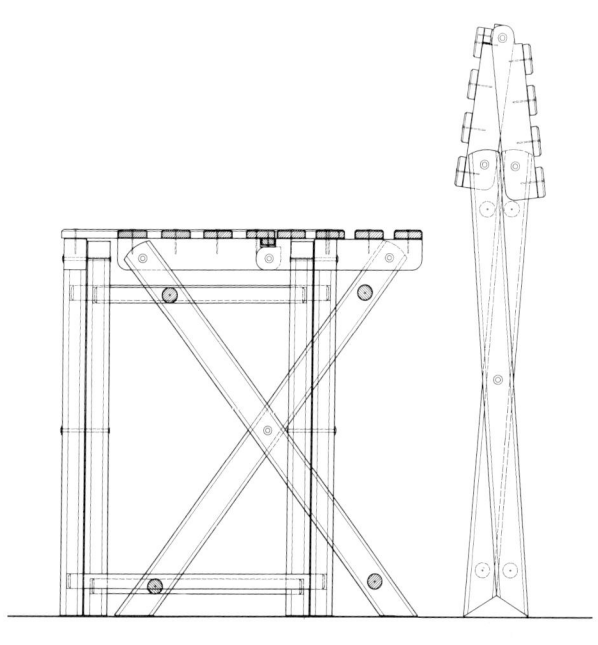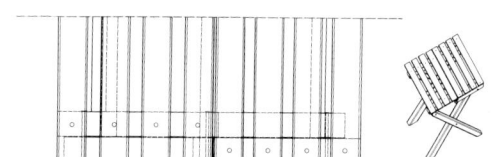

5
Feldstockerl
Thonet Österreich (Feldstockerl Nr. 6/6856). Der quadratische Sitz hat Rohrgeflecht-Bespannung und ist nach unten klappbar. Die Sperre erfolgt durch einen Gurtstreifen an der Unterseite; gedrechselte Füsse aus Buche

Sitzhöhe	41 cm
Sitzfläche	42 x 35 cm
Gefalten	54 cm

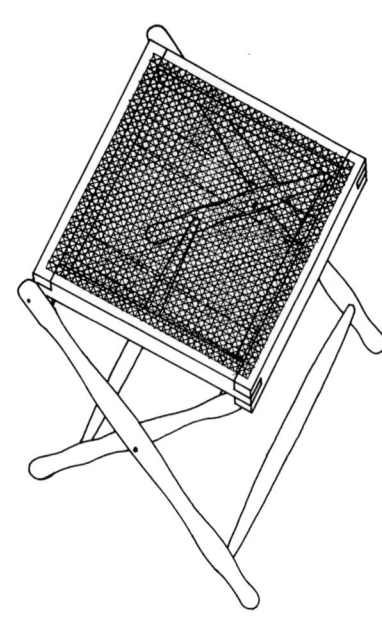

5
Camp stool
Thonet Austria (camp stool No. 6/6856). The square seat is cane covered and folds down. A strap on the underside secures the seat; turned legs of beech.

Height of seat	41 cm
Area of seat	42 x 35 cm
Folded	54 cm

36

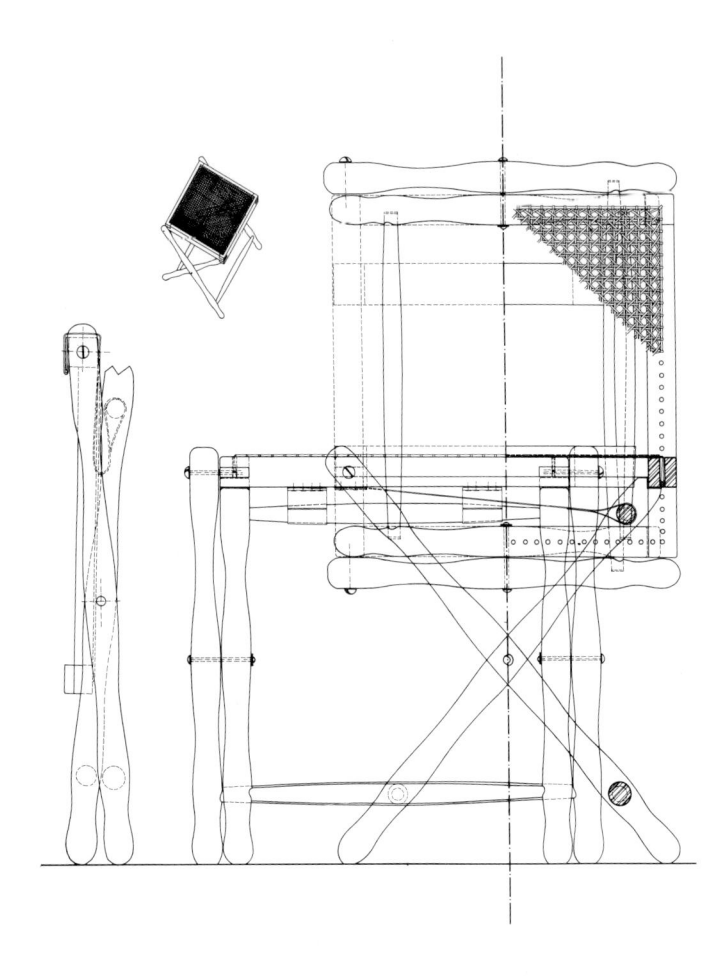

6
Tragstockerl
aus Rundstäben mit Bambusimitation und Stoffsitz.
Die Arretierung wird durch einen
Stab mit Beschlag auf die unteren
Sprossen wirkend entsprechend
hergestellt. Im zusammen-
klappbaren Zustand durch
2 Bügel tragbar.

Sitzhöhe	50 cm
Sitzfläche	50 x 23 cm
geklappt mit Griff	75 cm

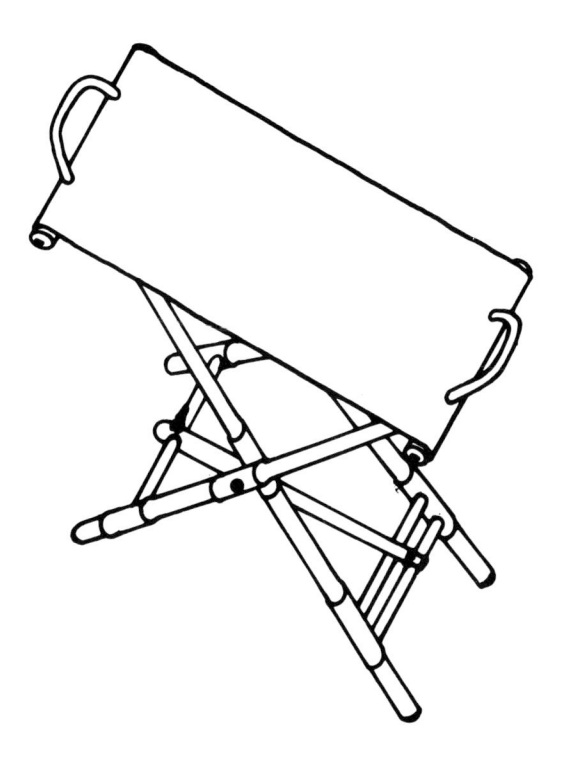

6
Portable stool
imitation bamboo rods and fabric seat. The lock is provided by a metal-fitted rod engaging on the lower stretchers. Carried by two handles when collapsed.
Height of seat   50 cm
Area of seat   50 x 23 cm
Folded with handle   75 cm

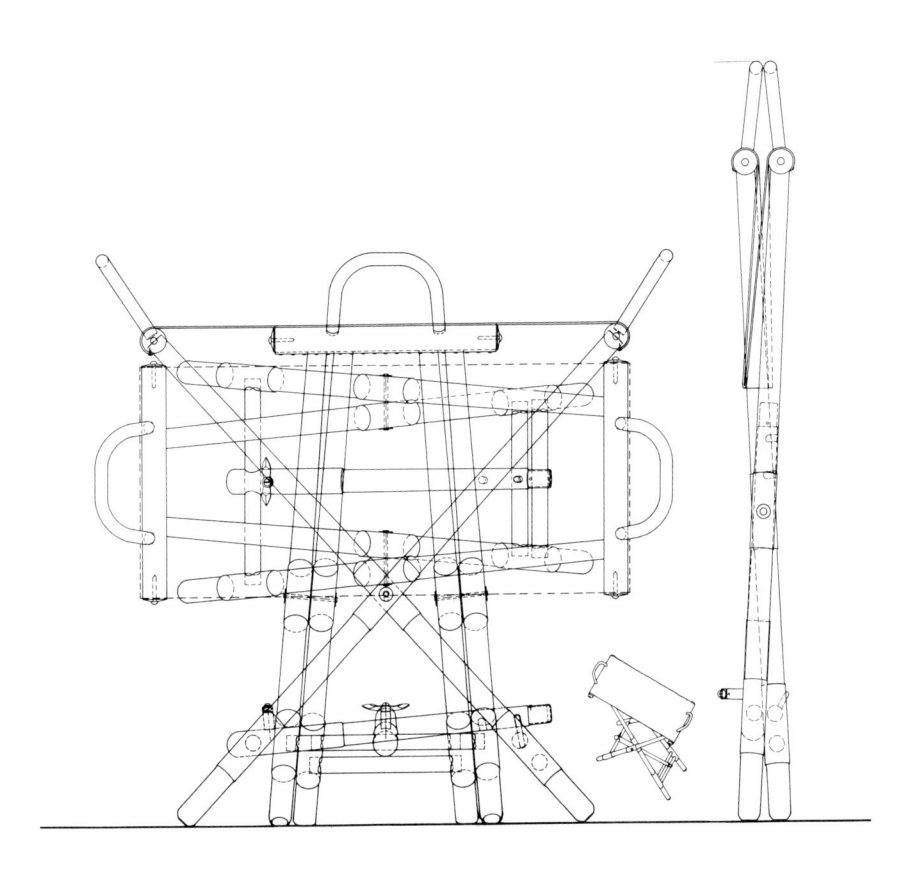

7
Feldstockerl
Thonet (Feldstockerl Nr. 1/6871)
aus Bugholz mit gedrechseltem
Achsholz und Schrauben-
verbindung sowie Stoffsitz
Sitzhöhe   48 cm
Sitzfläche   47 x 45 cm
Geklappt   66 cm

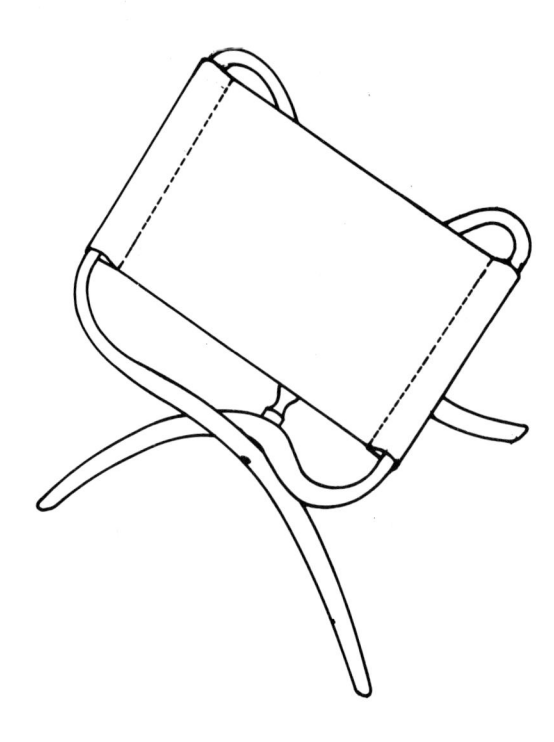

7
Camp stool
Thonet (camp stool No. 1/6871)
of bentwood with turned stretcher
and screwed joint, cloth seat.
Height of seat   48 cm
Area of seat   47 x 45 cm
Folded   66 cm

42

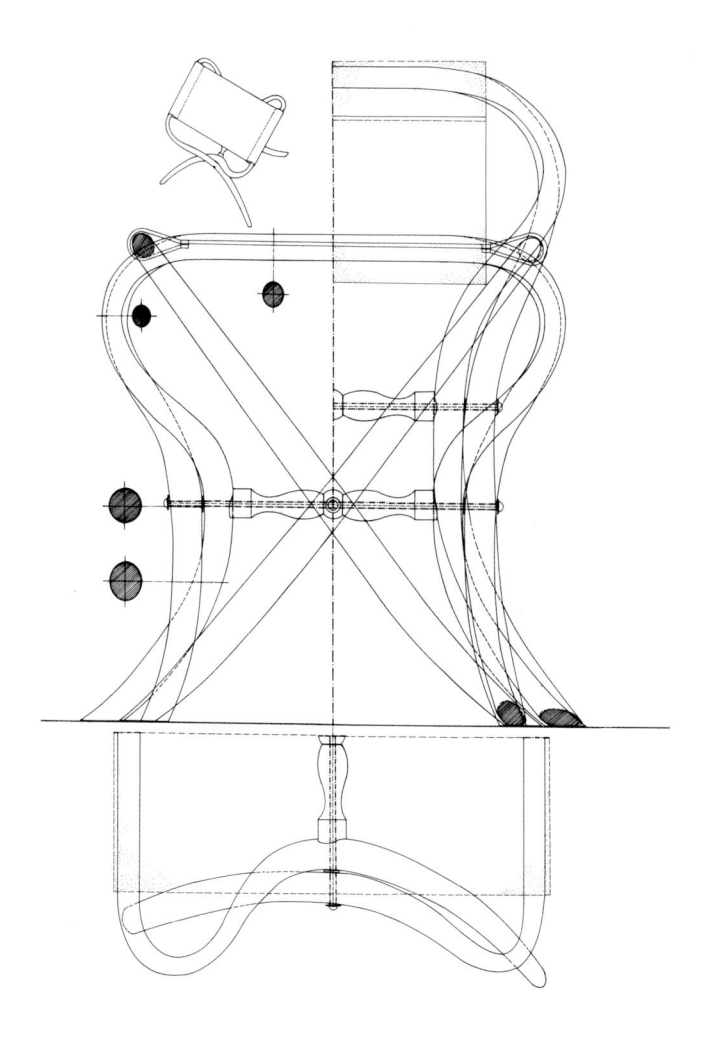

8
Kinderhocker
aus Bambusbrettchen zu einem
flachen Paket zusammenschieb-
und klappbar
Sitzhöhe   20 cm
Sitzfläche   24 x 20 cm
gefalten   29 cm

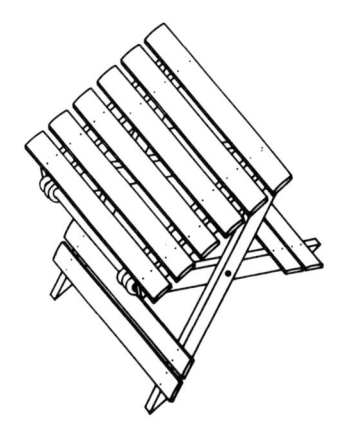

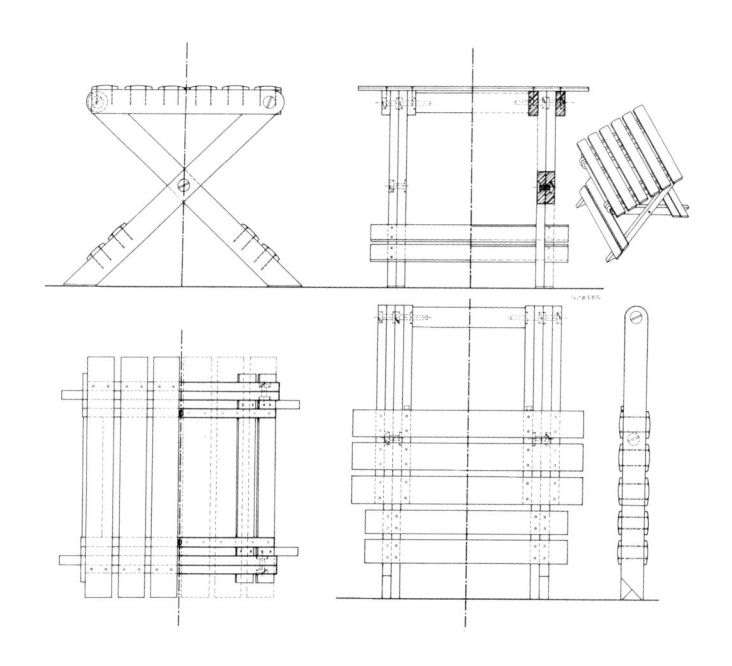

8
Child's stool
of bamboo slats, telescoping and folding flat
Height of seat 20 cm
Area of seat 24 x 20 cm
Folded 29 cm 44

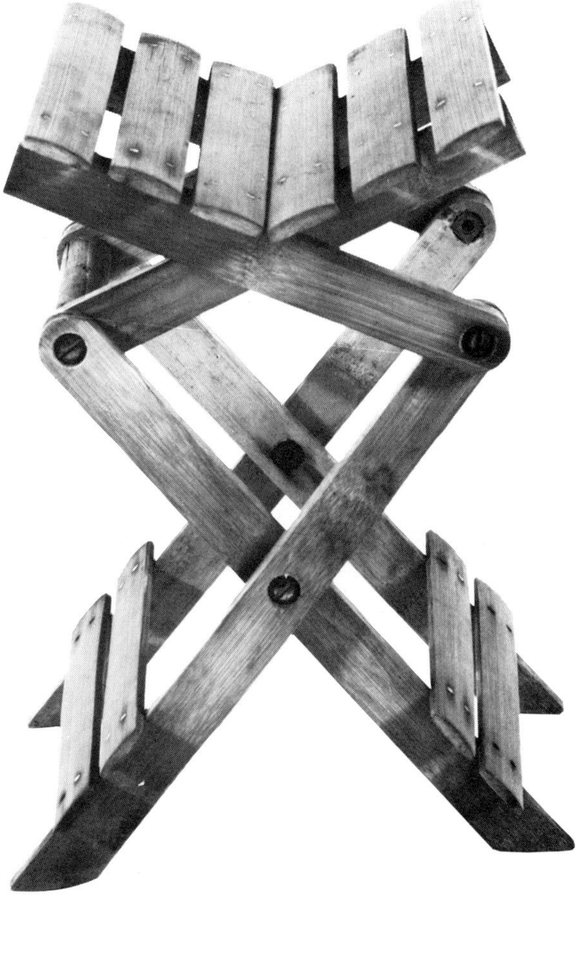

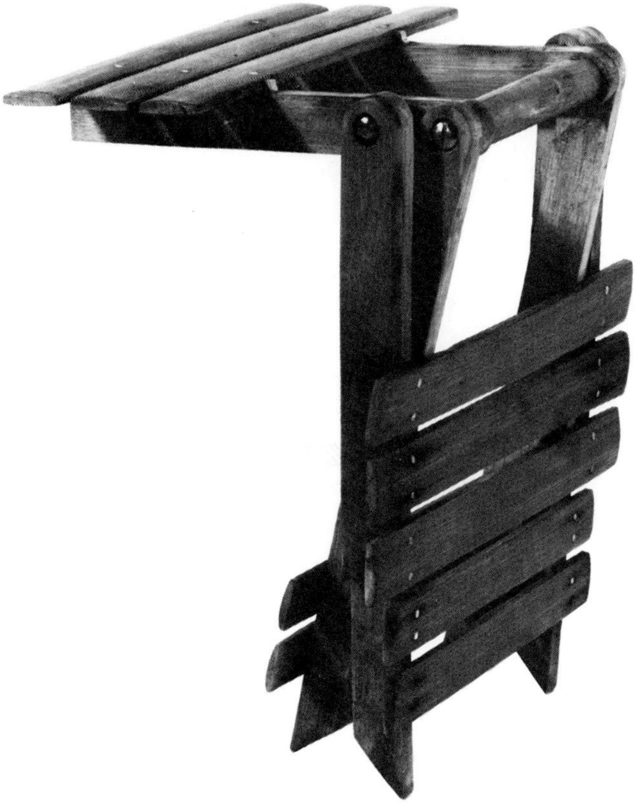

9
Kinderklappsessel
aus Buche mit nach oben klappbarer Sitzfläche aus gelochtem Sperrholz. Die Umlenkung erfolgt hebelartig mit einem Metallsteg.

Sitzhöhe	34 cm
Sitzfläche	26 x 23 cm
Gesamthöhe	65 cm
(71 cm geklappt)	

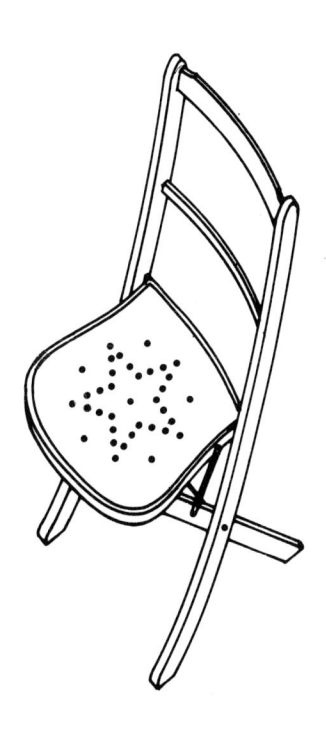

9
Child's folding chair
of beech with a fold-up seat of
perforated plywood. Lever system
with metal link.
Height of seat 34 cm
Area of seat 26 x 23 cm
Overall height 65 cm
(71 cm folded)

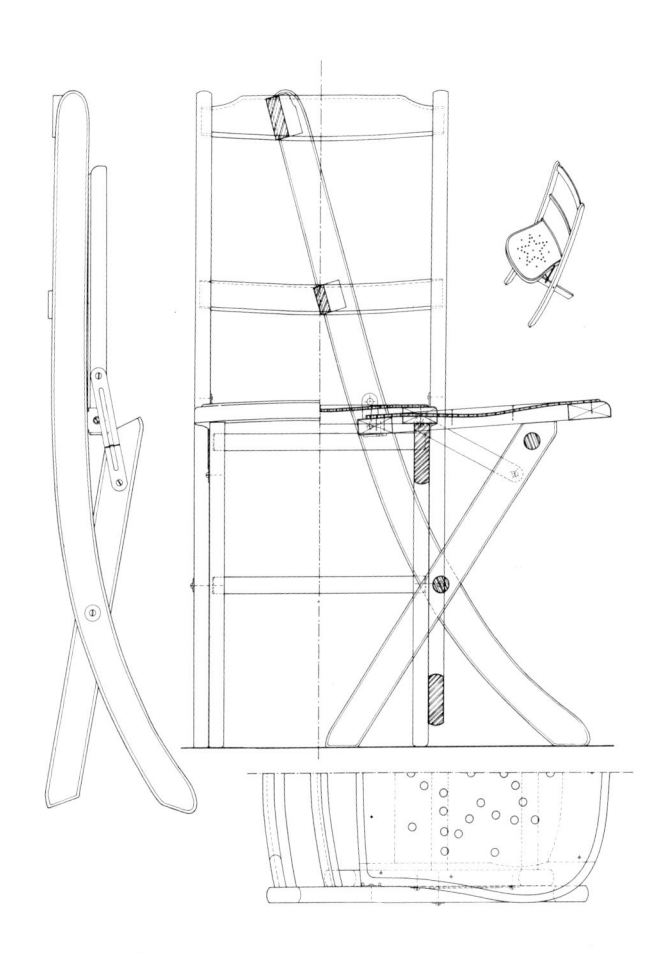

10
Kinderklappstuhl
mit Brettlsitz aus Buche. Der Lehnenteil ist nach vorne klappbar. Die Führung des Sitzteils erfolgt über Rundstahlhebel.

Sitzhöhe	32 cm
Sitzfläche	29 x 24 cm
Gesamthöhe	56 cm
	(64 cm)

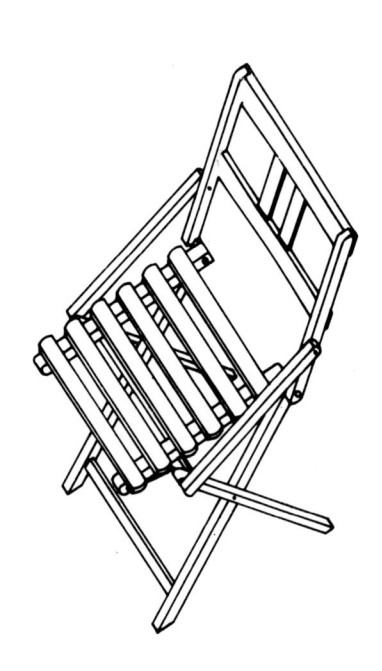

10
Child's folding chair
with slat seat of beech. The back folds forward. The seat is guided by a steel rod lever.
Height of seat 32 cm
Area of seat 29 x 24 cm
Overall height 56 cm
(64 cm)

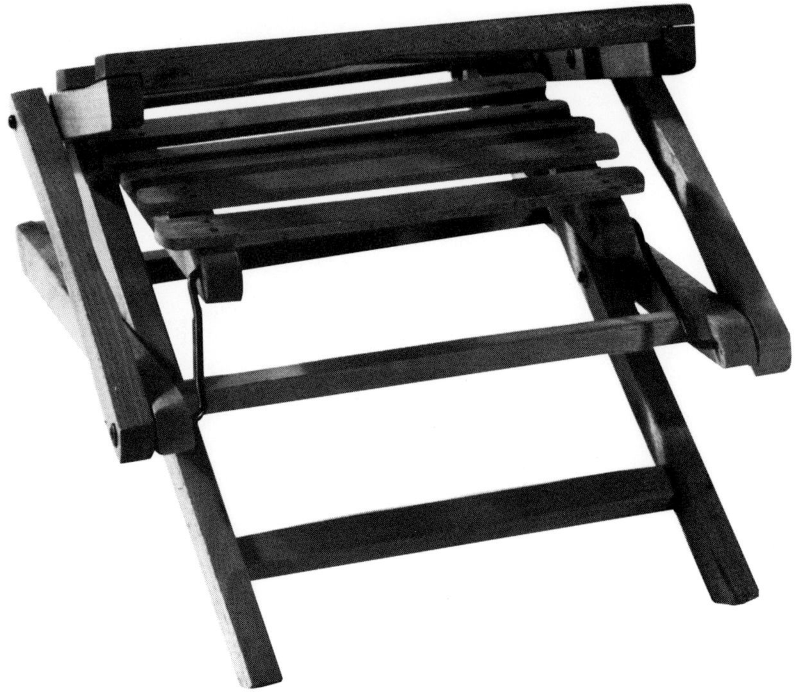

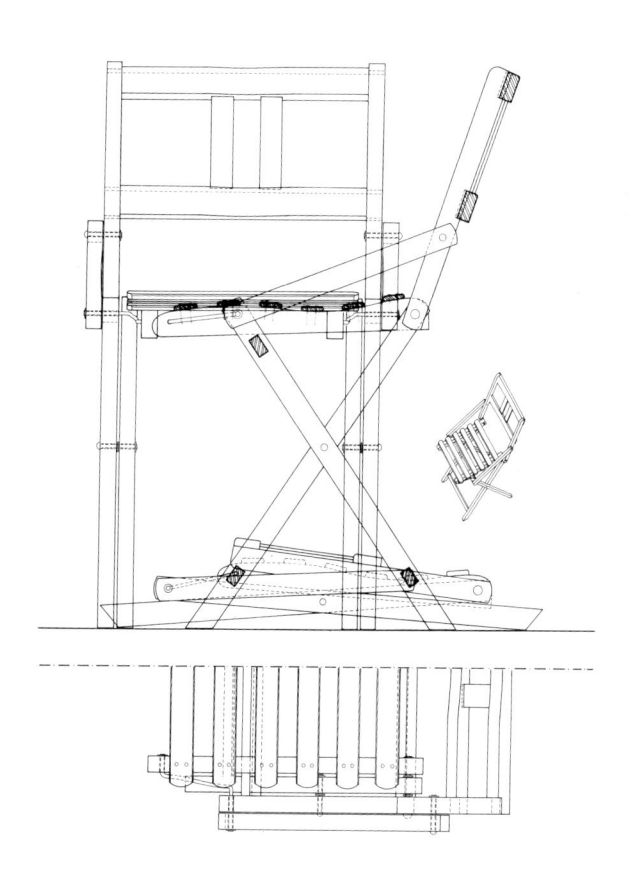

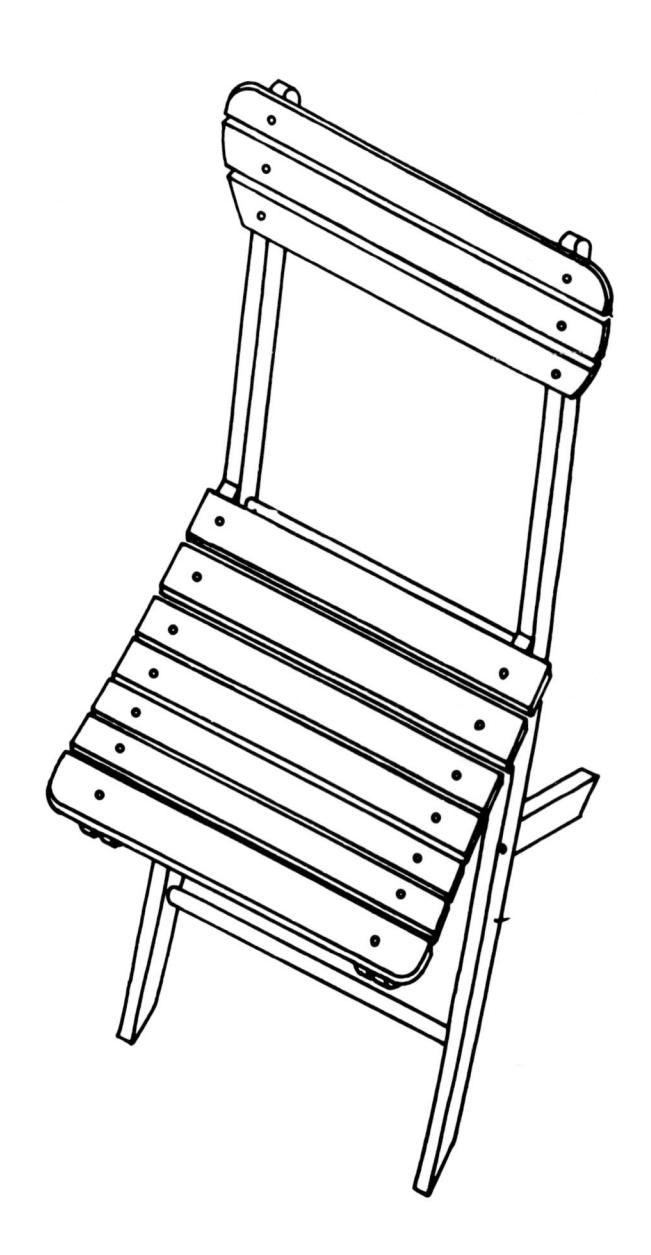

11
Gartensessel
aus Vierkanthölzern mit von
rückwärts aufklappbarem Brettlsitz. Die Querverbindungen sind
durch Rundstäbe hergestellt.
Sitzhöhe   45 cm
Sitzfläche   41 x 36 cm
Gesamthöhe   82 cm
                    (97 cm)

11
Garden chair
of squared wood with slat seat
tipping up backwards. The
stretchers are rods.
Height of seat   45 cm
Area of seat   41 x 36 cm
Overall height   82 cm
                    (97 cm)

56

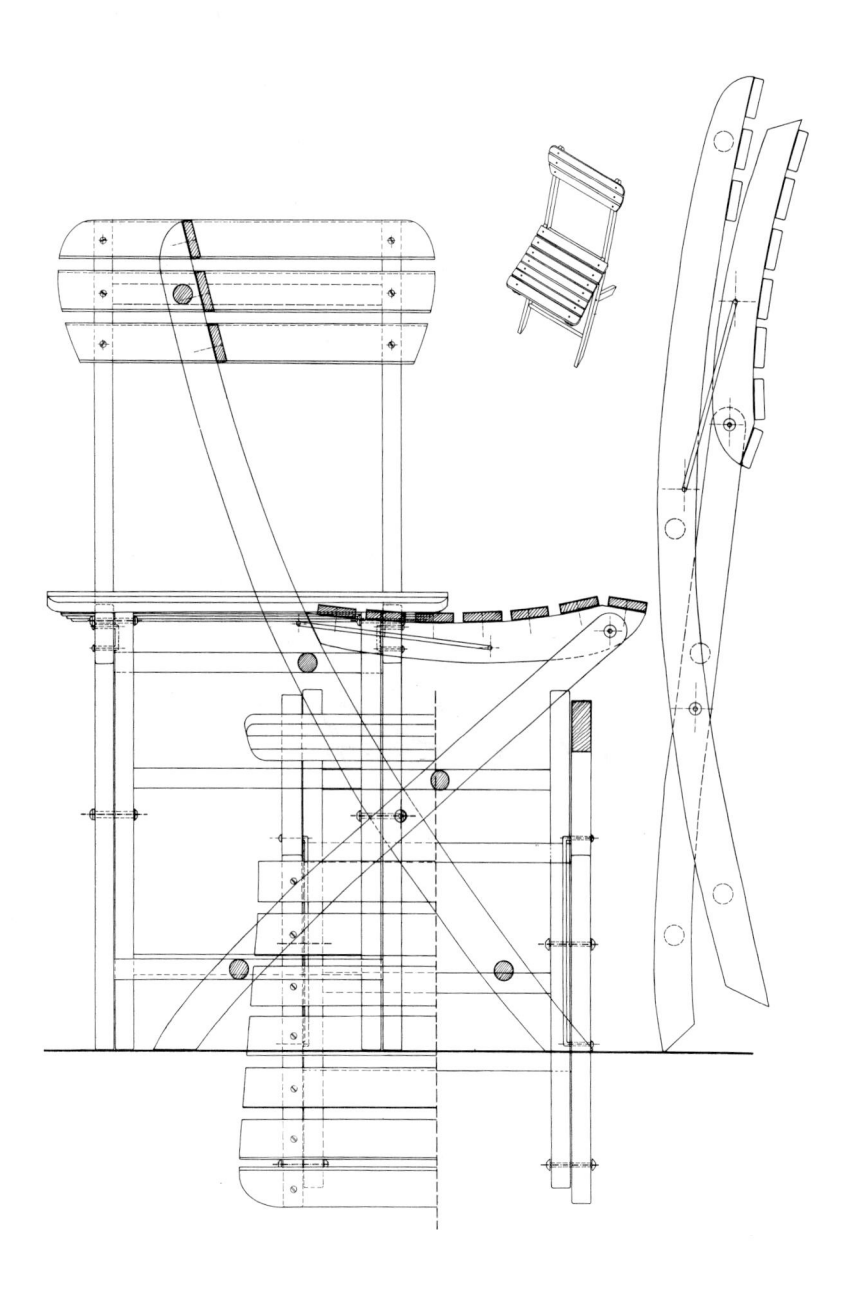

12
Schiffsessel
mit Sitzführung in den äusseren
Beinschenkeln mit Brettlsitz und
Brettllehne, die mitgeschwenkt
wird.

Sitzhöhe	45 cm
Sitzfläche	34 x 34 cm
Gesamthöhe	81 cm
	(97 cm)

12
Boat chair
with seat running in grooves in
the outer legs, with slat seat and
slat back pivoting simultaneously.
Height of seat 45 cm
Area of seat 34 x 34 cm
Overall height 81 cm
(97 cm)

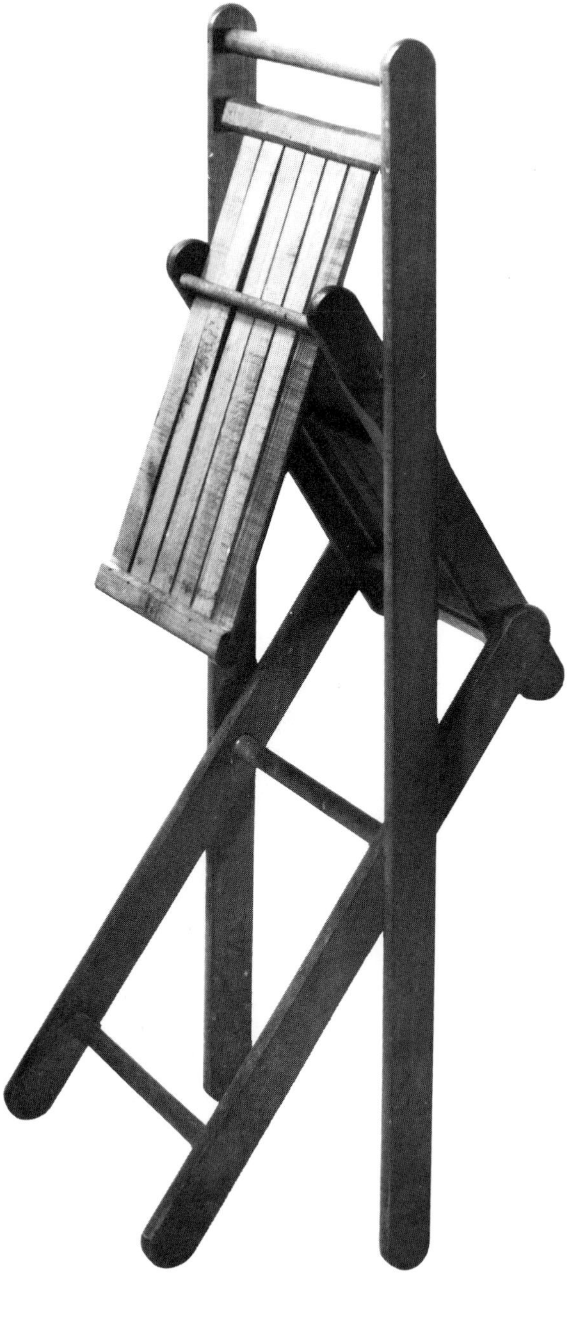

60

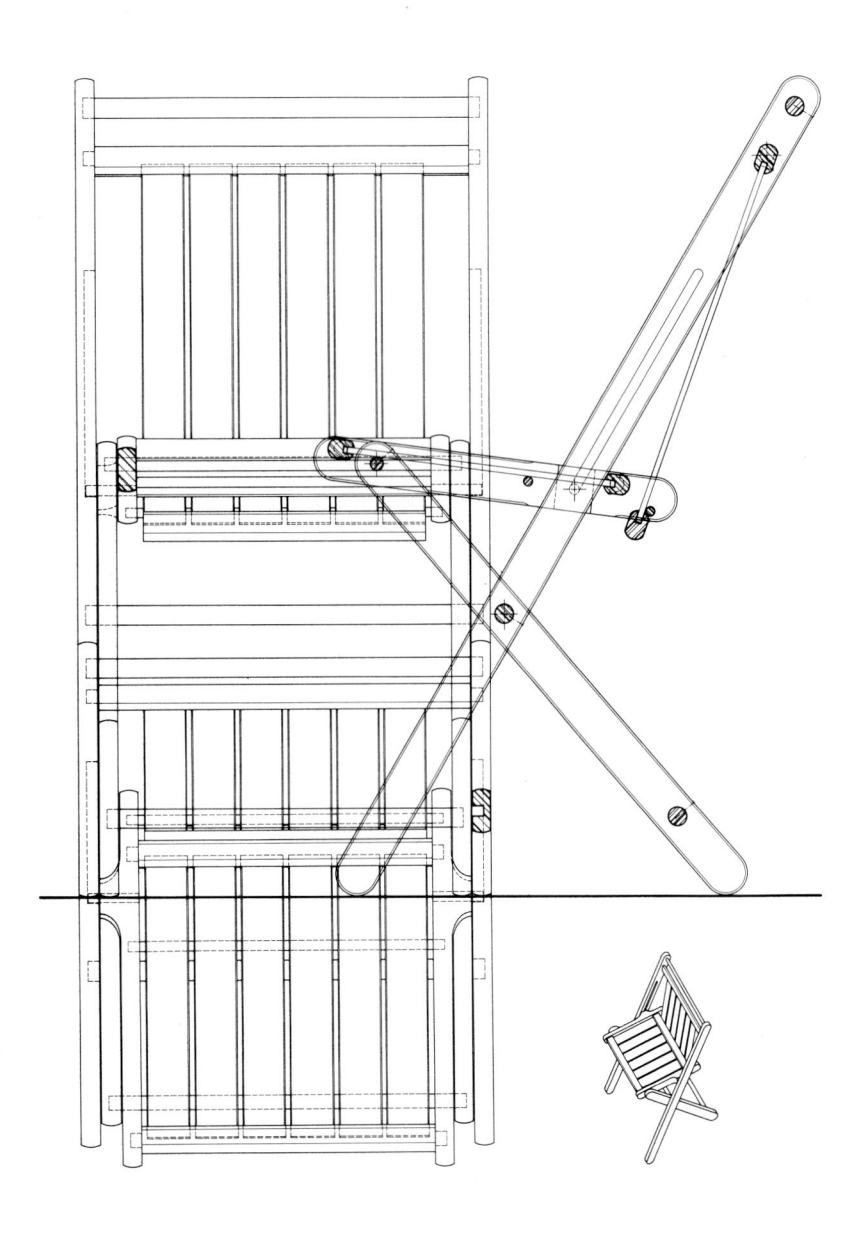

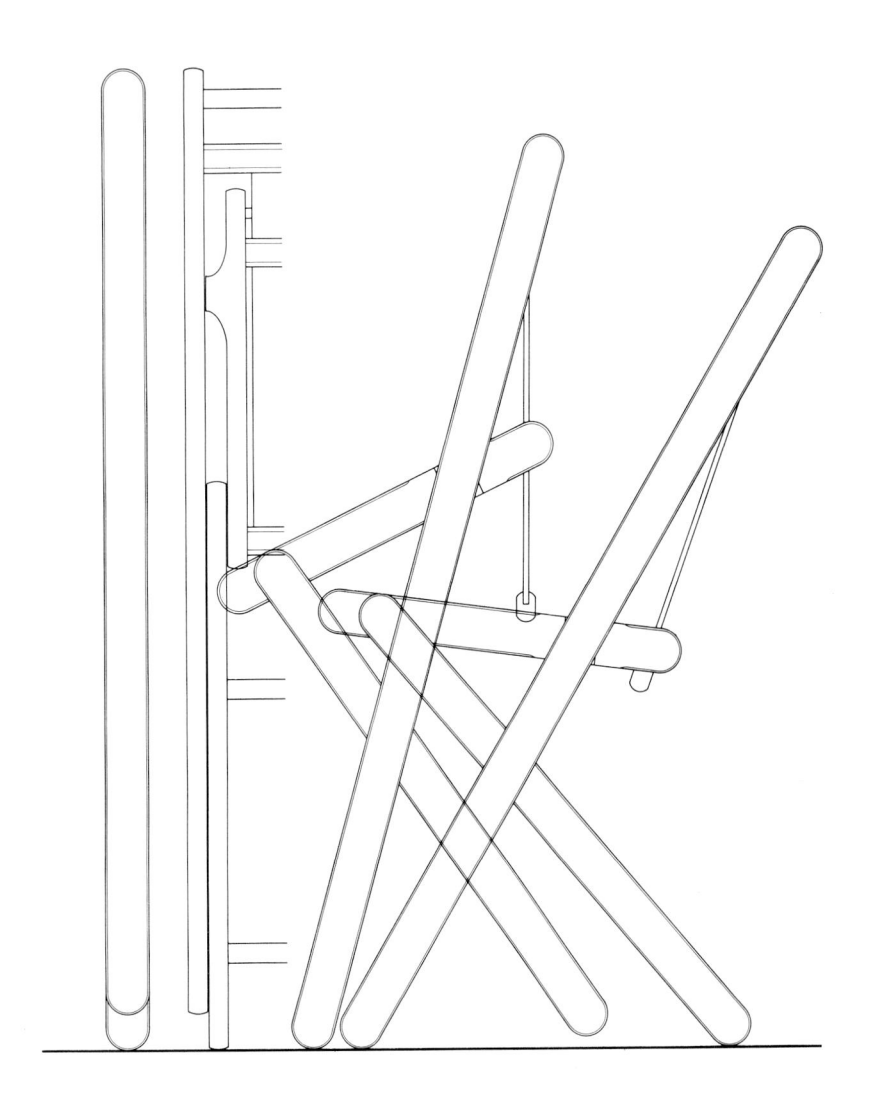

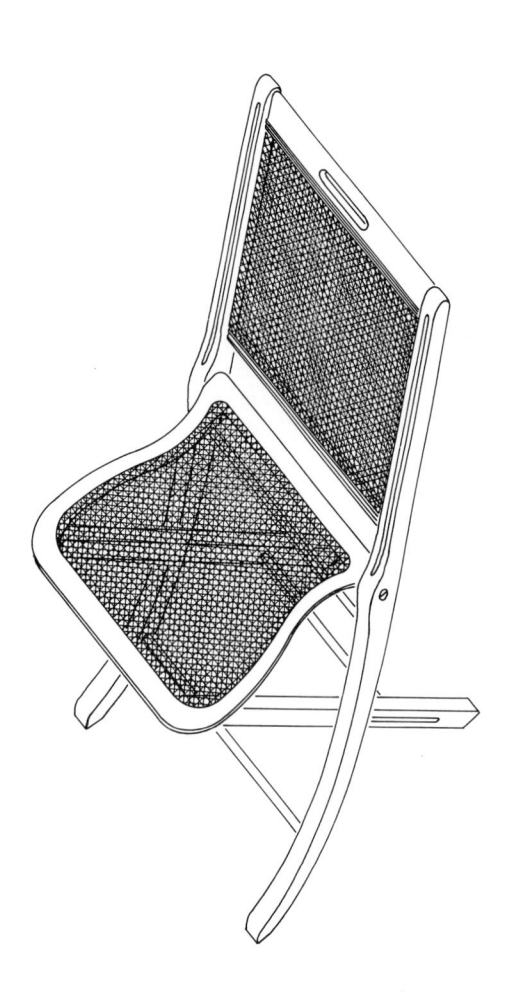

13
Schiffsessel
dunkelrot gebeizt mit nach oben von vorne klappbarem Sitz. Die rückwärtigen Füsse sind zur Führung geschlitzt. Sitz und Lehne sind mit Rohrgeflecht bespannt.

Sitzhöhe	47 cm
Sitzfläche	38 x 34 cm
Gesamthöhe	83 cm
	(95 cm)

13
Boat chair
stained dark red with seat folding up from the front. The rear legs are grooved as guides. Cane seat and back.

Height of seat	47 cm
Area of seat	38 x 34 cm
Overall height	83 cm
	(95 cm)

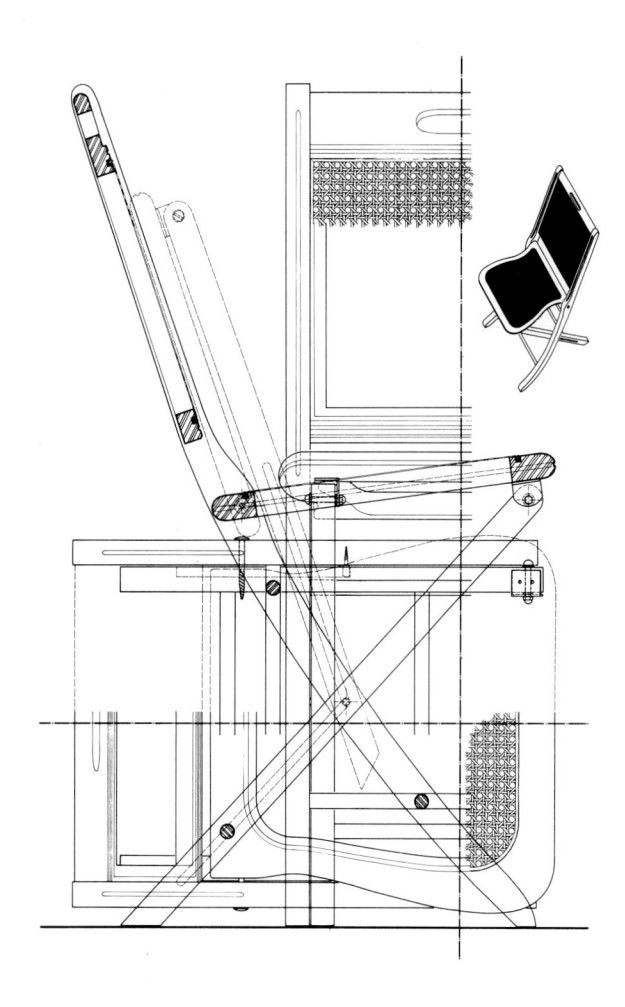

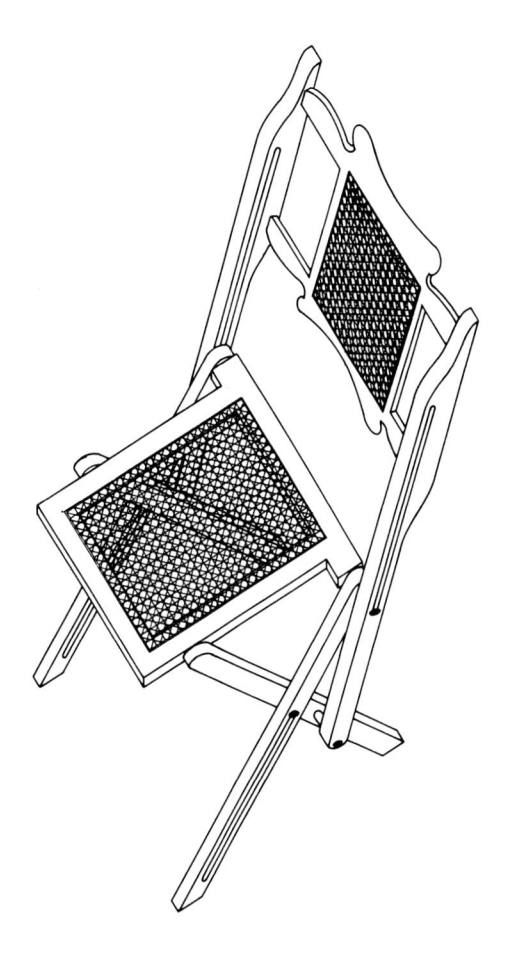

14
Englischer Klappsessel
aus Eiche mit von hinten nach
oben hochklappbarem Sitz. Die
vorderen Füsse sind zur Führung
geschlitzt ebenso die Rückenlehne. Sitz und Lehne sind mit
Rohrgeflecht bespannt.

Sitzhöhe	41 cm
Sitzfläche	27 x 34 cm
Gesamthöhe	83 cm
	(93 cm)

14
English folding chair
of oak with seat tipping up from behind. The front legs are slotted as guides, likewise the back. Cane seat and back.
Height of seat   41 cm
Area of seat   27 x 34 cm
Overall height   83 cm
    (93 cm)

67

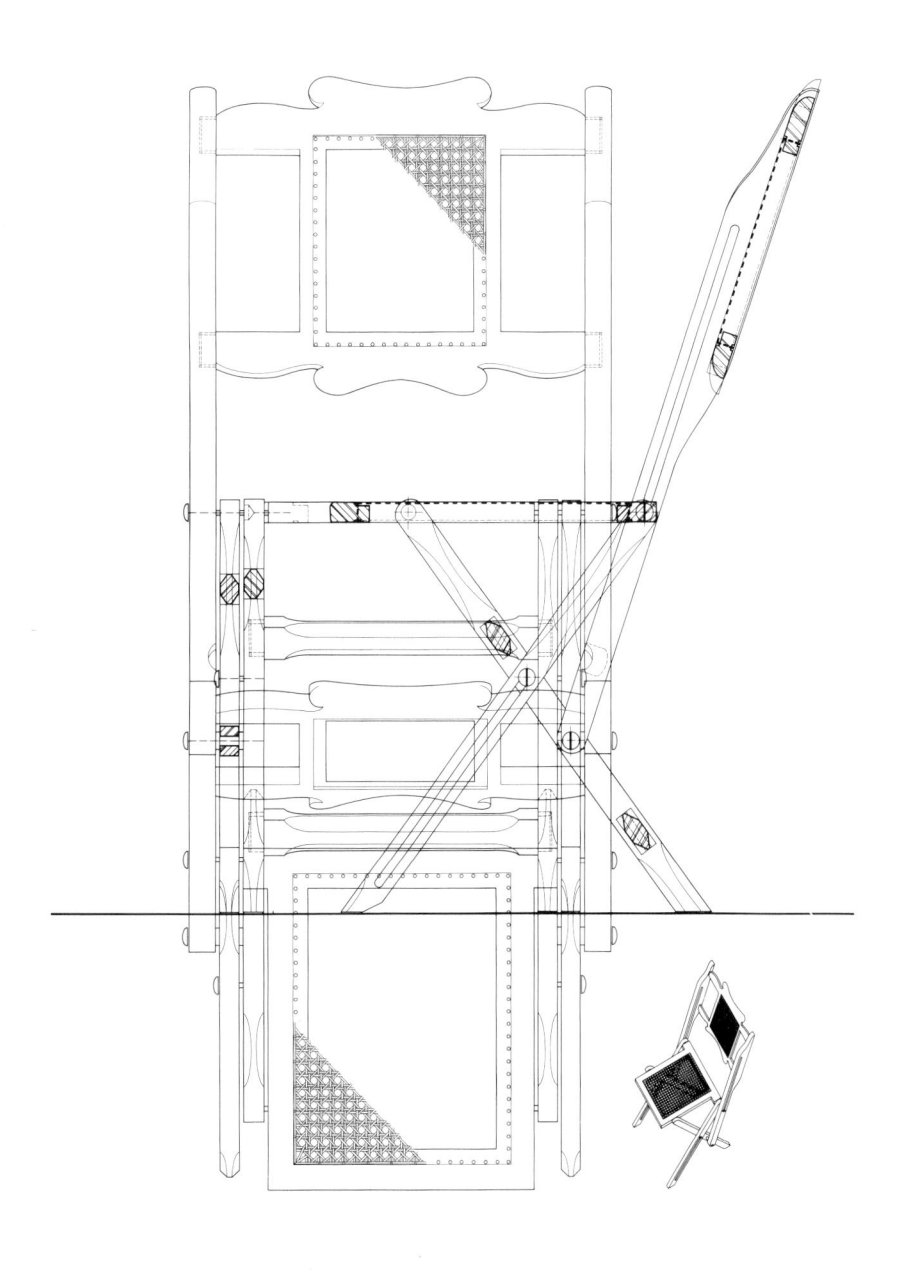

68

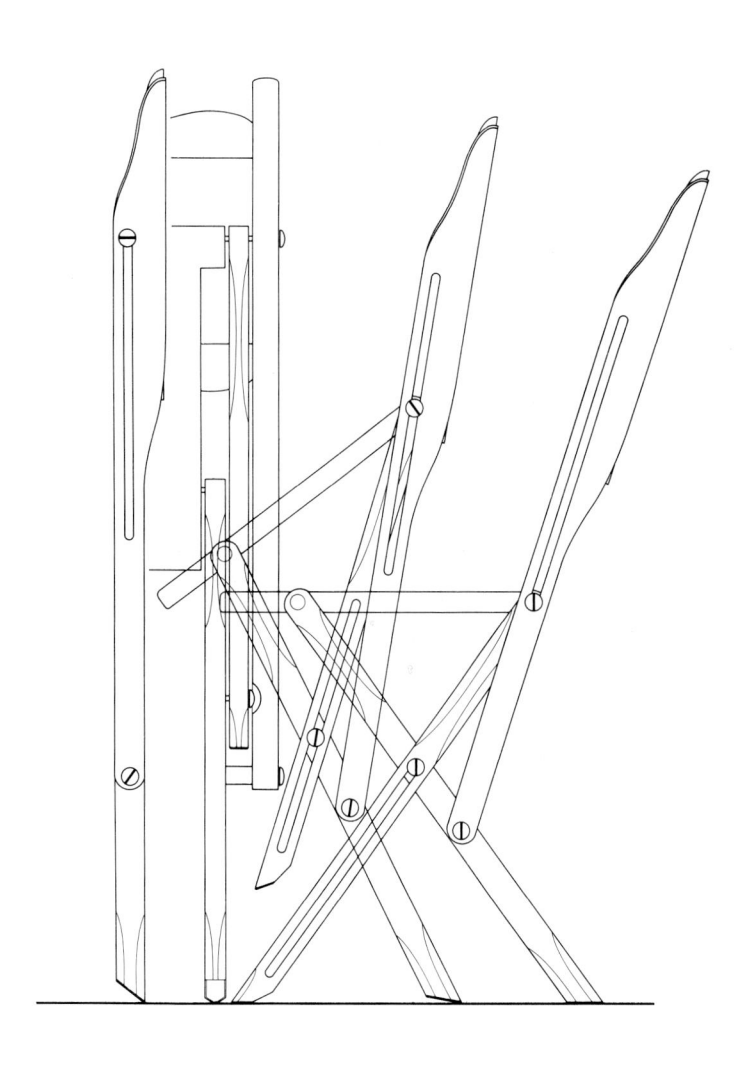

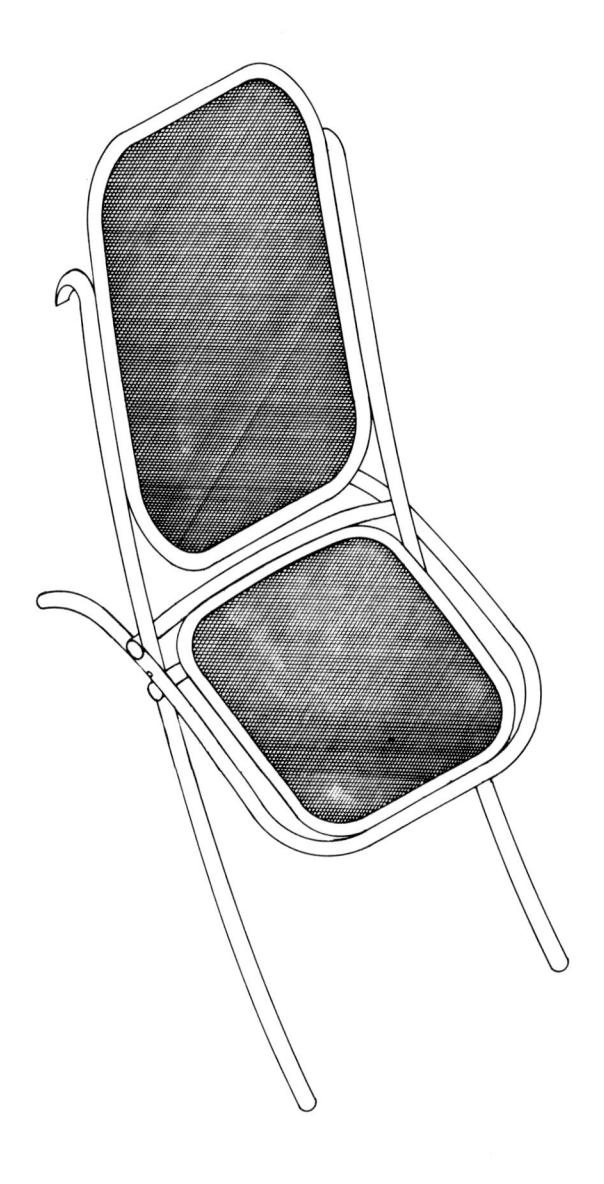

15
Kaminsessel
Thonet (Nr. 1/6311), zusammenlegbarer Sitz und Lehnteil. Die Sperre erfolgt durch Querstege in den beiden Teilen. Bugholz mit Rohrgeflechtverspannung.

Sitzhöhe	43 cm
Sitzfläche	46 x 47 cm
Gesamthöhe	85 cm
	(116 cm)

15
Fireside chair
Thonet (No. 1/6311), collapsible seat and back. The lock is effected by crosspieces in both parts. Cane and bentwood.
Height of seat 43 cm
Area of seat 46 x 47 cm
Overall height 85 cm
(116 cm)

71

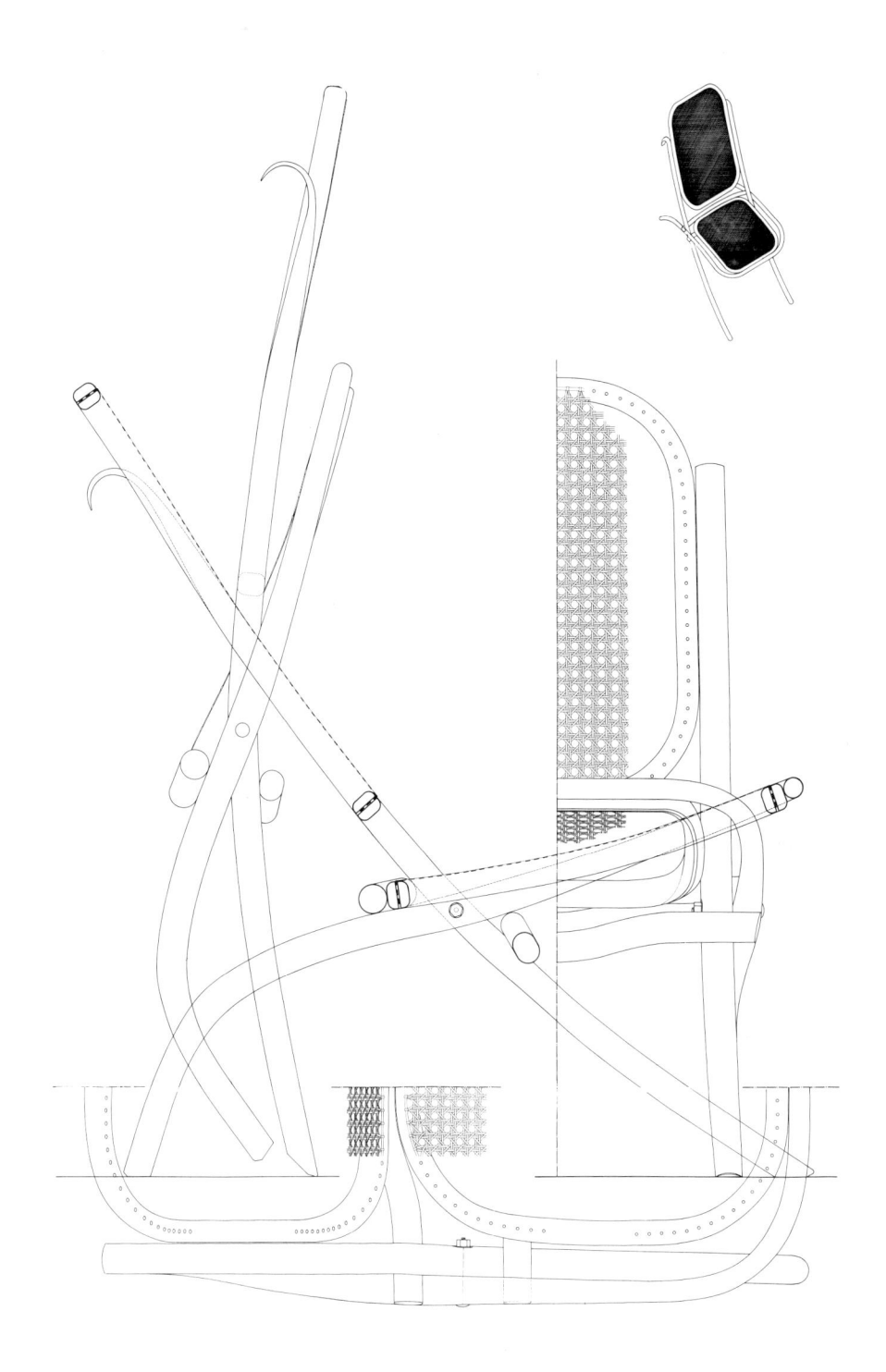

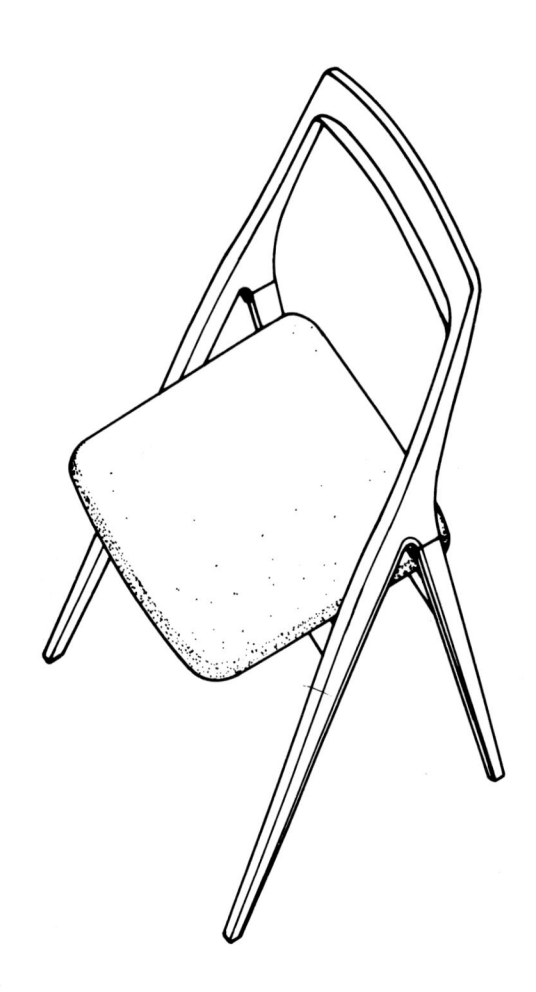

16
Holzsessel
mit gepolstertem Sitz (Hochschule für angewandte Kunst).
Die rückwärtigen Füsse werden zu den Vorderfüssen und der Sitz von vorne hochgeklappt.
Material Teakholz.

Sitzhöhe	45 cm
Sitzfläche	47 x 42 cm
Gesamthöhe	77 cm
	(88 cm)

16
Wooden chair
with upholstered seat (University of Applied Arts). The back legs are folded up to the front legs and the seat raised from the front.
Material: teak
Height of seat   45 cm
Area of seat   47 x 42 cm
Overall height   77 cm
                      (88 cm)

74

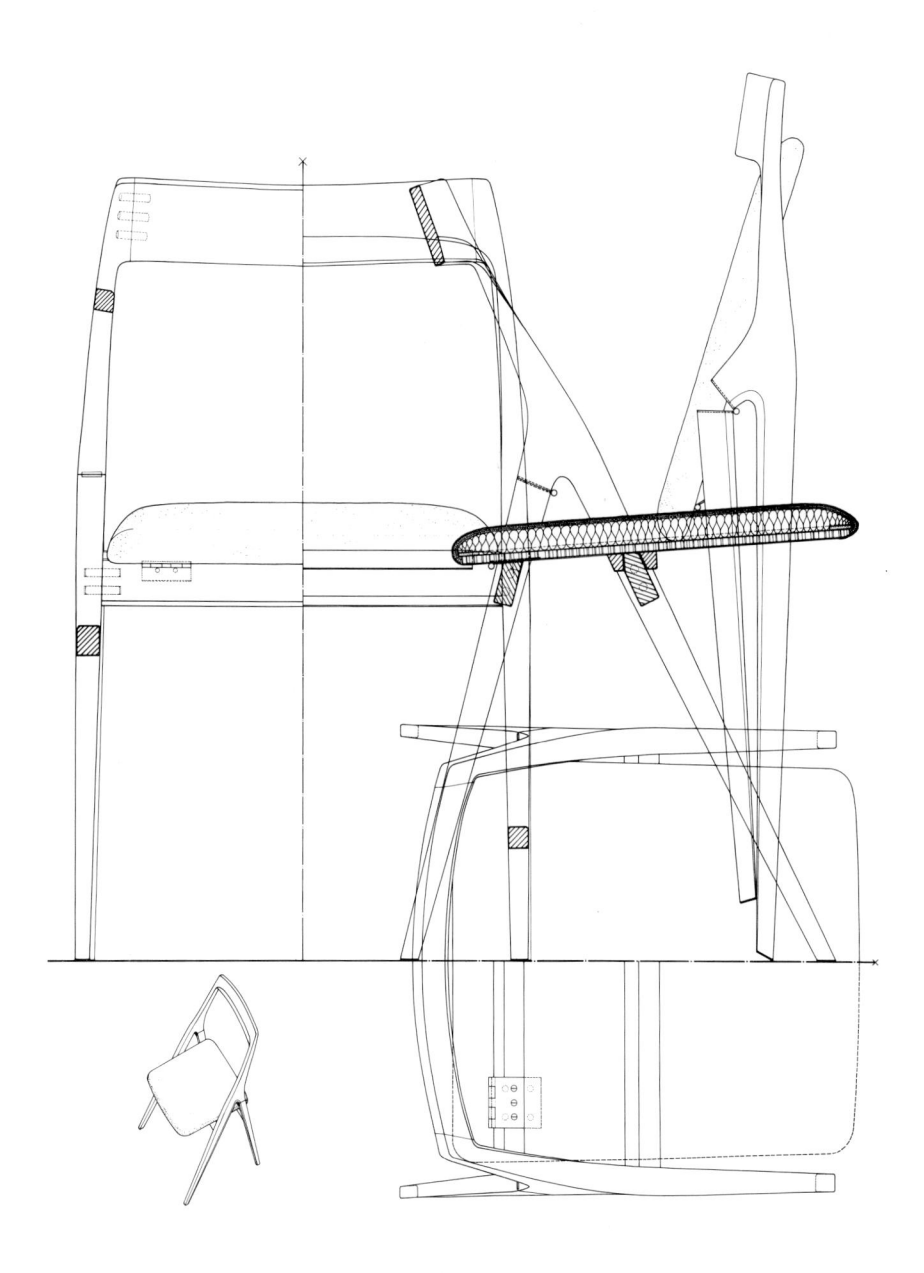

17
Klappstuhl ‚Plia'
(Italien 1969) aus Stahlrohr verchromt, Sitz und Lehne aus durchsichtigem Plexiglas.
Sitzhöhe 45 cm
Sitzfläche 46 x 42 cm
Gesamthöhe 76 cm
(87 cm)

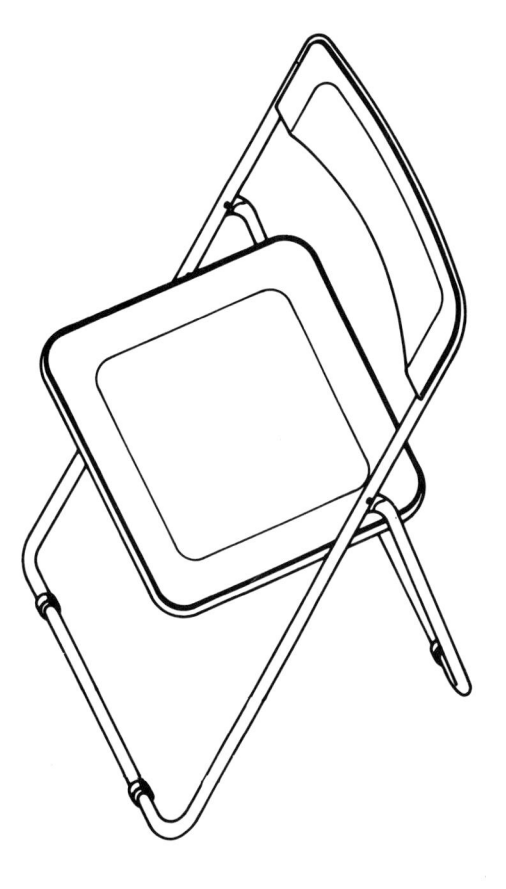

17
'Plia' folding chair
(Italy 1969) of chromium-plated
tubular steel, seat and back of
transparent Plexiglas.
Height of seat 45 cm
Area of seat 46 x 42 cm
Overall height 76 cm
(87 cm)

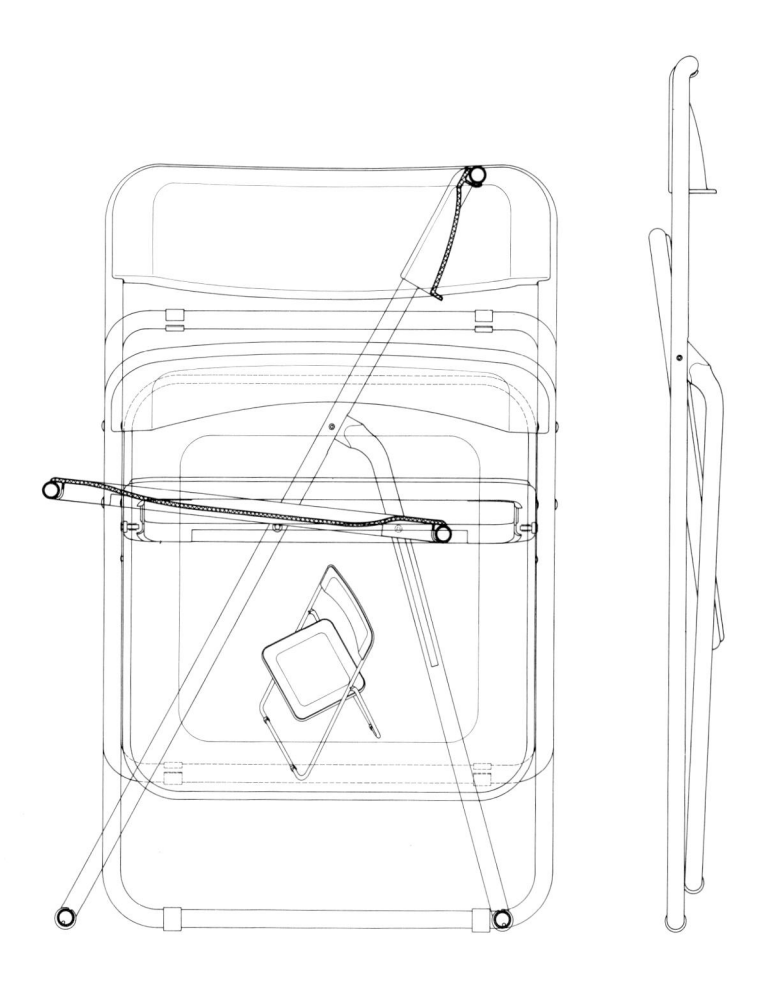

18
Damenschaukelstuhl
Die Verbindung von Sitz- und Lehnteil erfolgt über Stahlspangen. Buchenstäbe mit Stoffbespannung am Sitz und Sprossen am Lehnteil.

Sitzhöhe	38 cm
Sitzfläche	42 x 38 cm
Gesamthöhe	86 cm
	(104 cm)

18
Lady's rocking chair
The seat and back are joined by metal hinges. Seat of fabric-covered beech slats, spindle back.

Height of seat	38 cm
Area of seat	42 x 38 cm
Overall height	86 cm
	(104 cm)

80

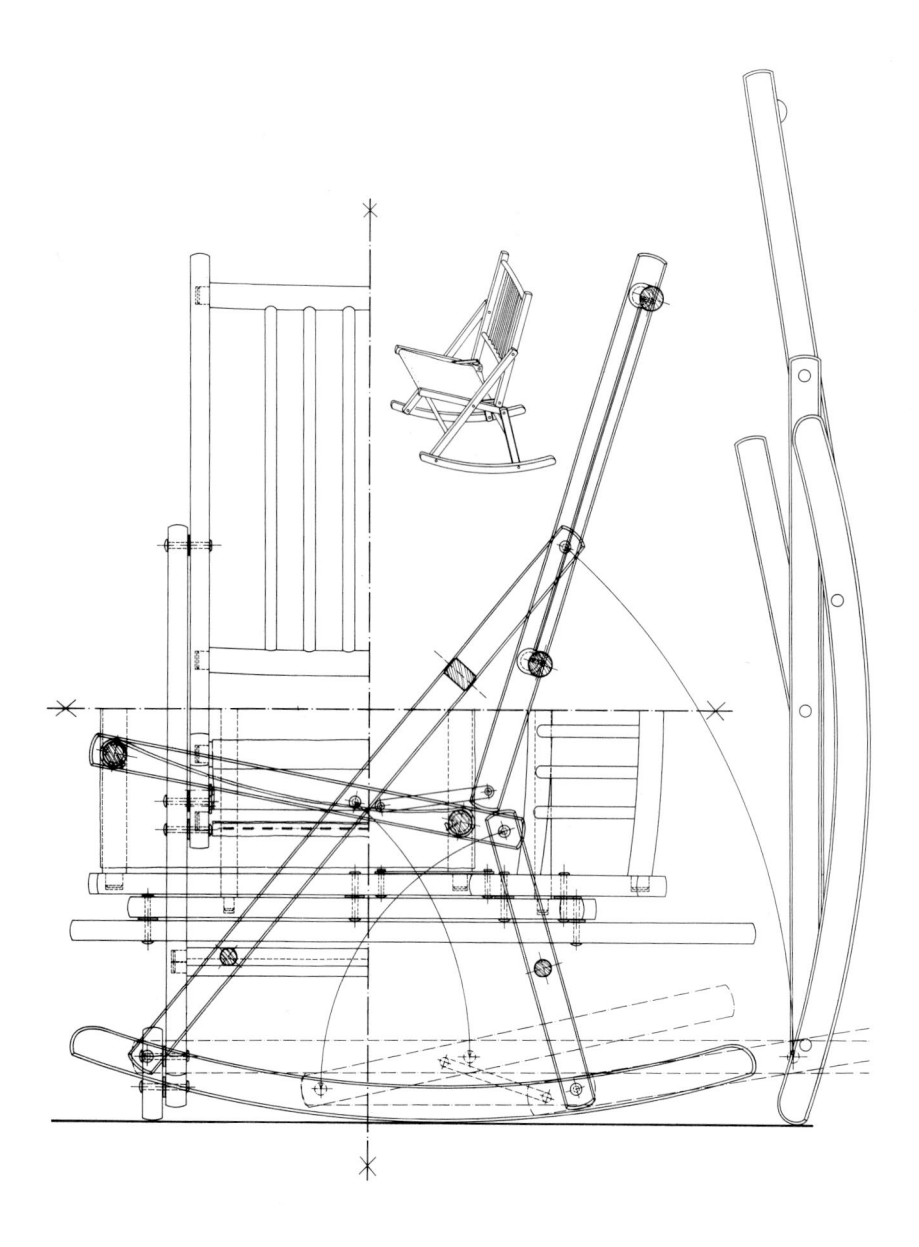

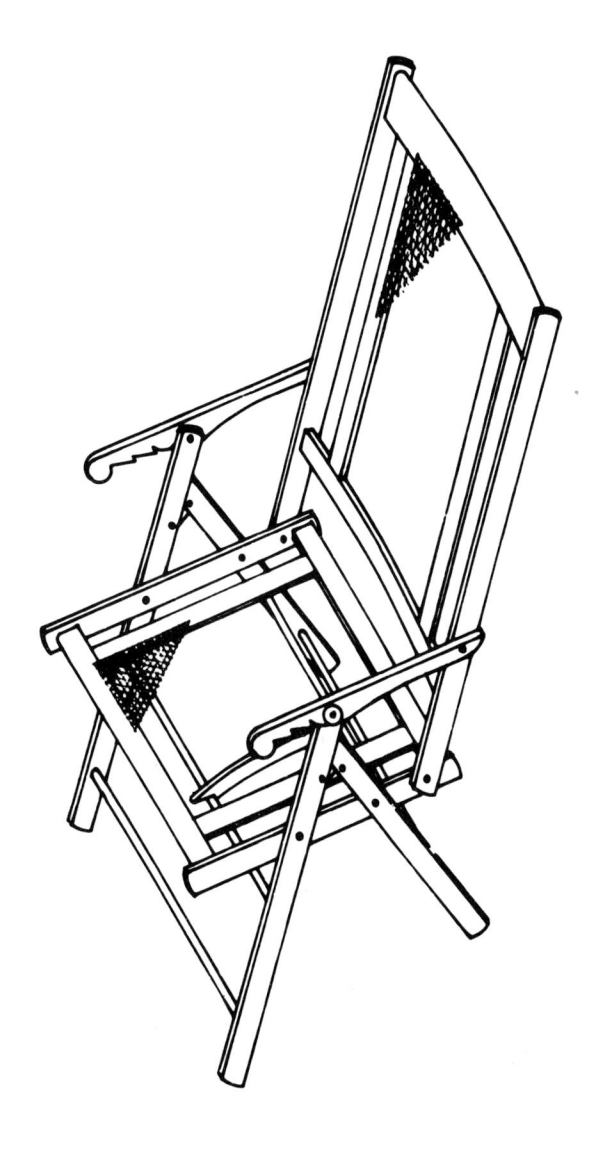

19
Verstellbarer Klapparmstuhl
(englisch).
Die Neigungsverstellung des
Lehnteils erfolgt über Heben und
Einrasten der Armlehne. Der Sitz
klappt nach oben, die Arretierung
wird durch einen beweglichen
Holzarm am unteren Gestell
erreicht. Sitz und Lehne haben
Rohrgeflechtbespannung.

Sitzhöhe	41 cm
Sitzfläche	51 x 50 cm
Gesamthöhe	86 cm
	(108 cm)

19
Adjustable folding armchair (English). The angle of the back is adjusted by lifting and clicking the arms into place. The seat tips up, the lock is provided by a movable wooden arm in the underframe. Cane back and seat.

Height of seat	41 cm
Area of seat	51 x 50 cm
Overall height	86 cm (108 cm)

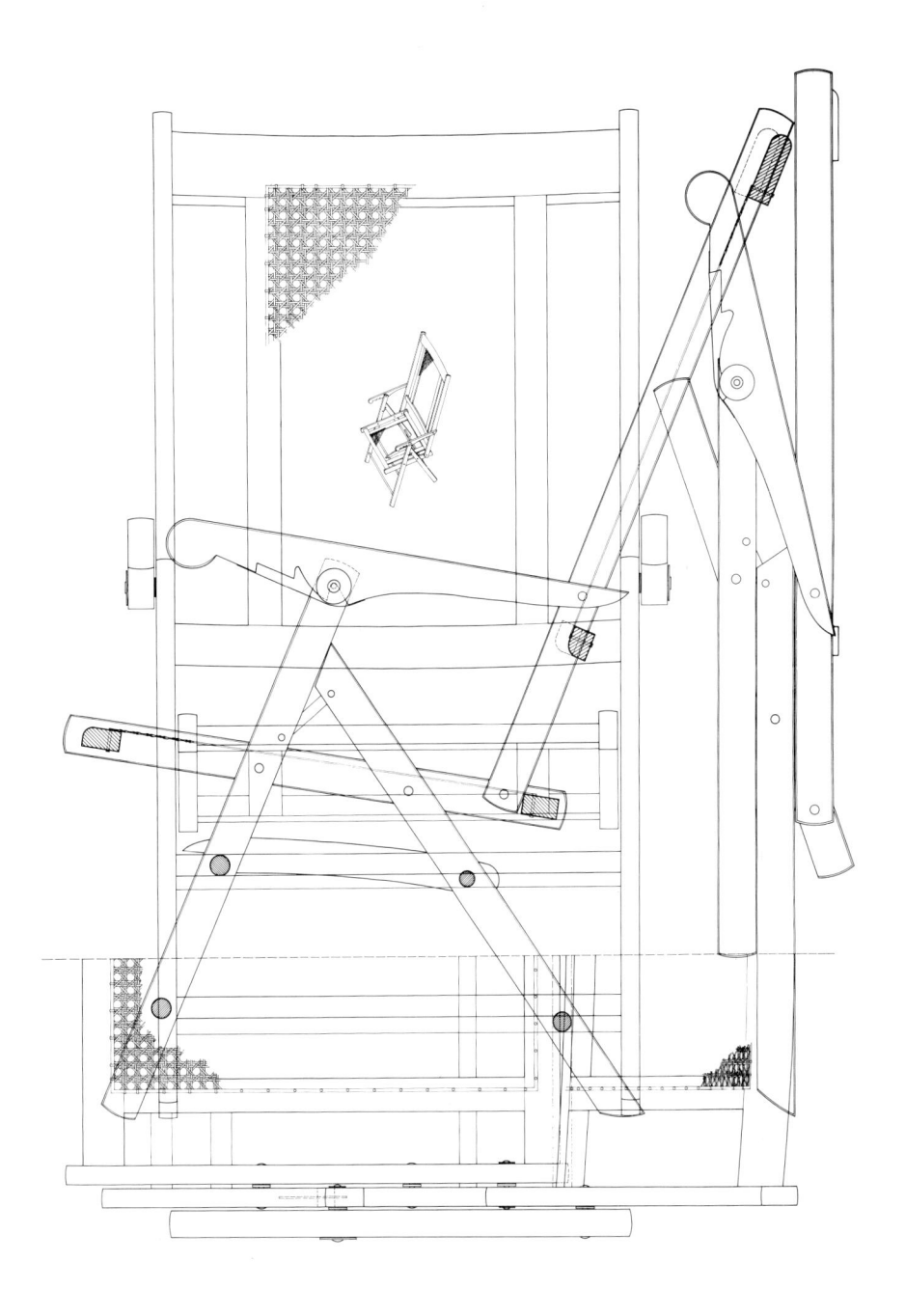

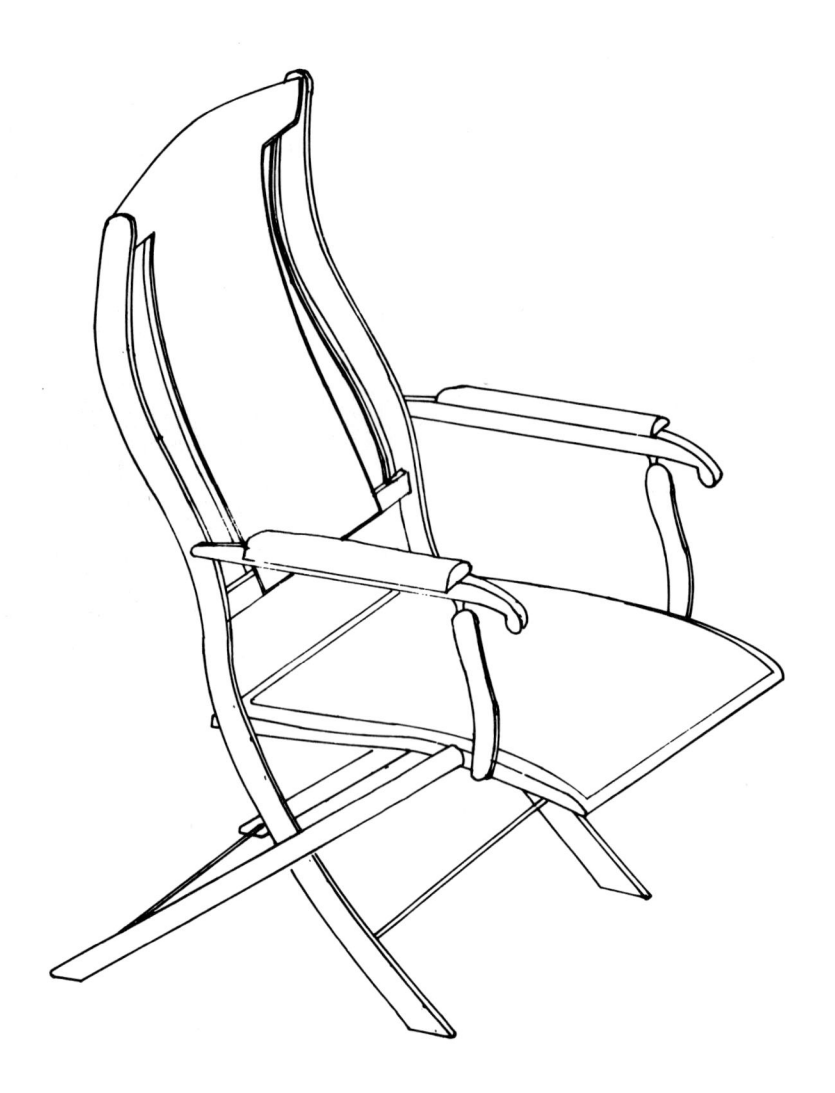

20
Klapparmstuhl
mit gepolstertem Sitz und Lehnteil und Armstützen (englisch). Die beiden äusseren Vorderfüsse sind zur Führung beim Klappen genutet. Der Sitz ist hochklappbar.

Sitzhöhe	40 cm
Sitzfläche	48 x 52 cm
Gesamthöhe	102 cm
	(120 cm)

20
Folding armchair
with upholstered seat, back and arms (English). The two front legs are grooved on the outside as a guiding device when the chair is folded.

Height of seat	40 cm
Area of seat	48 x 52 cm
Overall height	102 cm
	(120 cm)

86

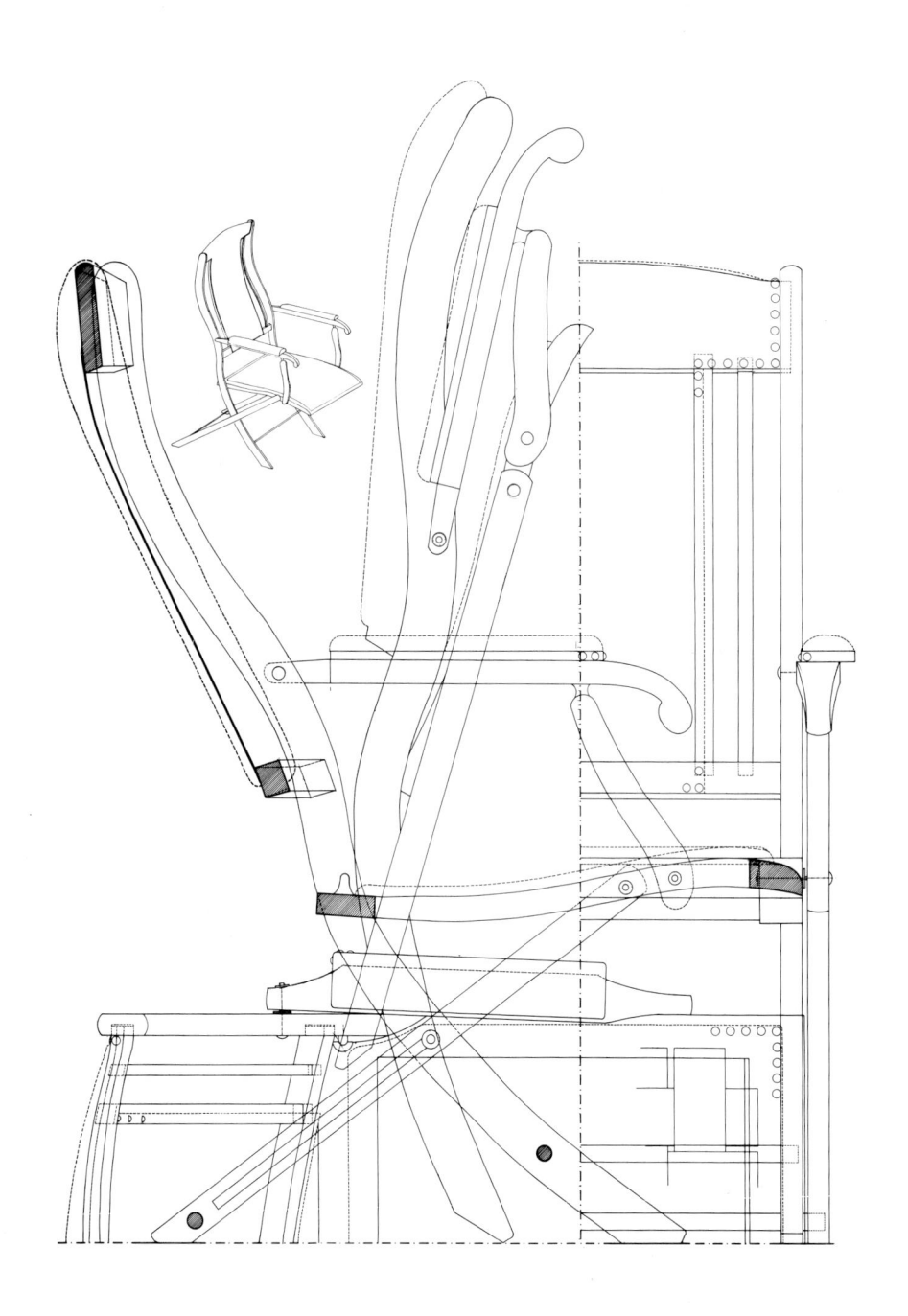

21
Regiesessel
mit beweglichem Rückenteil aus 2 mit x-förmigen Aussteifungen versehenen Rahmenteilen werden durch querliegende x-förmige Scheren, die für den Stoffsitz die Verspannung aufnehmen, gebildet. Die Sitzschere ist durch Metallbügel mit dem äusseren Rahmen verbunden.
Material: Buchenholz.

Sitzhöhe	40 cm
Sitzfläche	44 x 43 cm
Gesamthöhe	82 cm
	(69 cm)

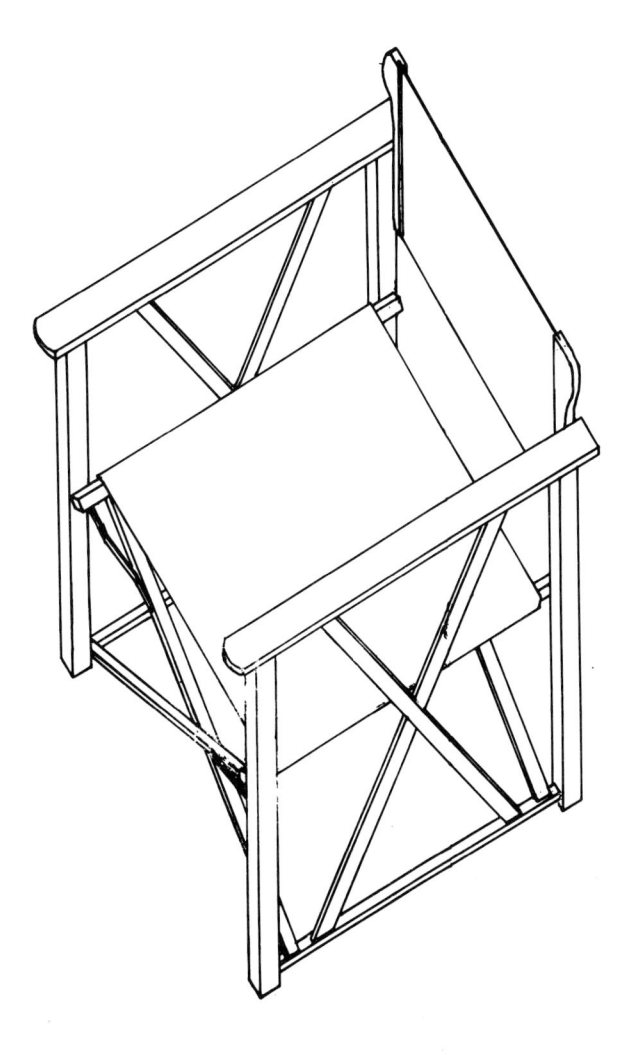

21
Film director's chair
with movable back. The 2 frames with X-shaped stays hold the bracing for the cloth seat. The seat scissors are attached to the outer frame with metal elements.
Material: beech
Height of seat 40 cm
Area of seat 44 x 43 cm
Overall height 82 cm
(69 cm)

88

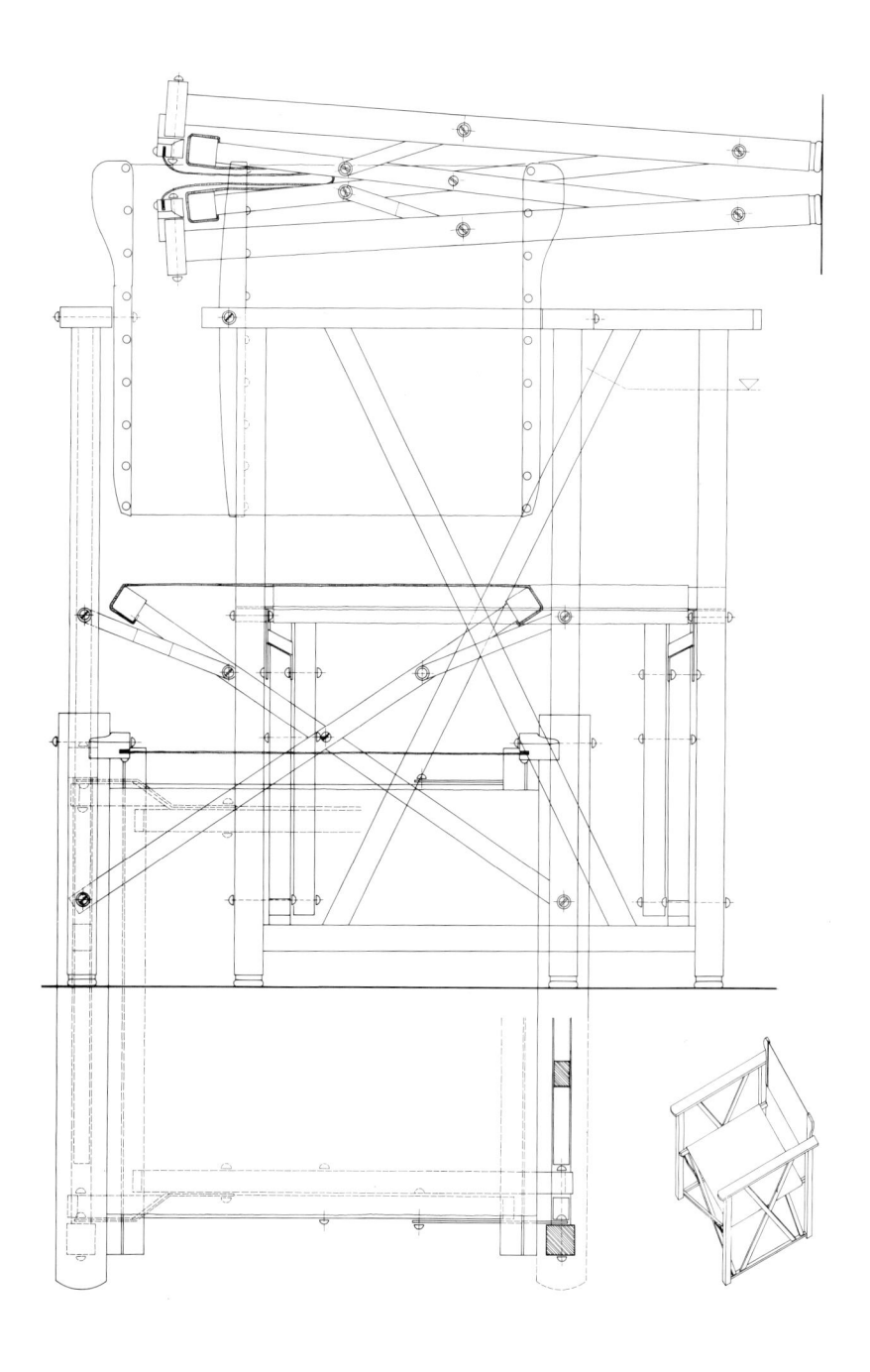

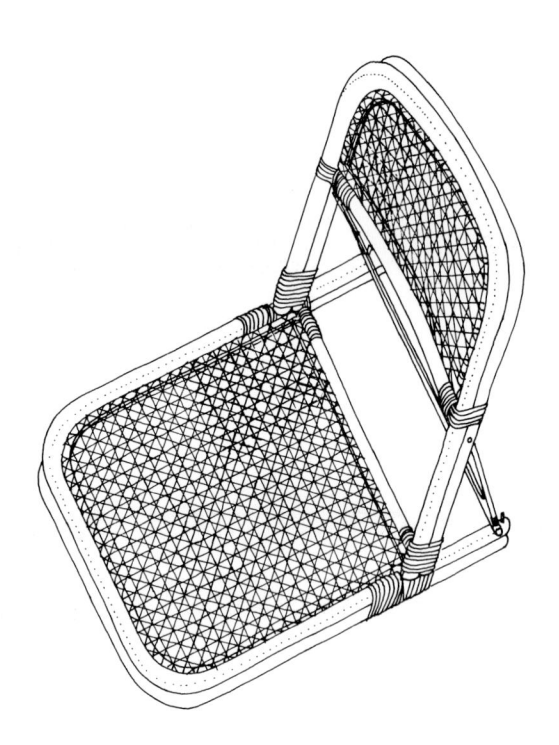

22
Bodensitz
aus verdoppeltem, gebogenen Peddigrohrrahmen mit Rohrgeflecht (Thailand). Der Lehnenteil wird nach vorne zusammengeklappt und hat ein u-förmiges bewegliches Stützteil, das in den verlängerten Sitzteil einrastet.
Rahmenhöhe 5½ cm
Sitzfläche 42 x 42 cm
Gesamthöhe 48 cm
(60 cm)

22
Floor seat
consisting of a double rattan cane frame with canework (Thailand). The back is folded forwards and has a movable U-shaped support which clicks into place on the lengthened seat.

Frame height	5½ cm
Area of seat	42 x 42 cm
Overall height	48 cm
	(60 cm)

92

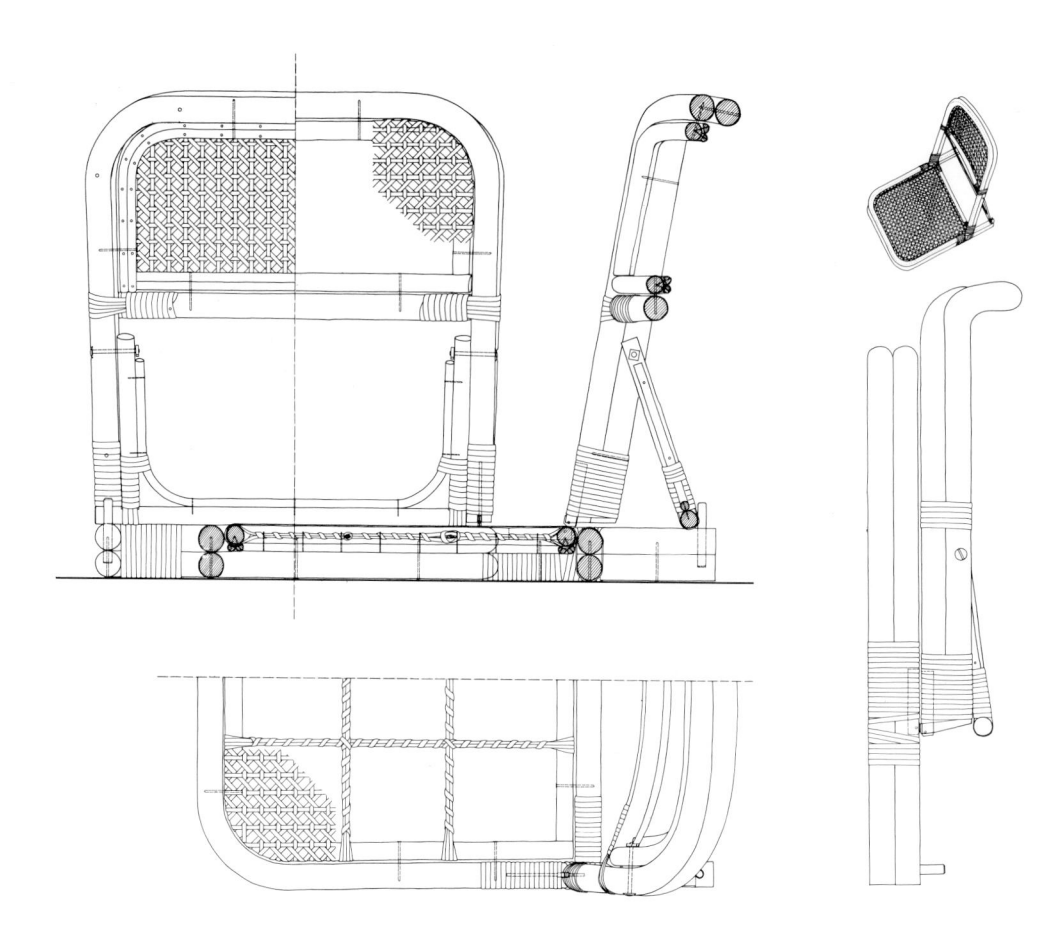

Nachwort des Herausgebers

Die vorliegende Arbeit, die leider nur kurzgefasste Einblicke in die Fülle der Klappstuhlmodelle zeigt, wäre undurchführbar gewesen, wenn nicht mein langjähriger Freund und Kollege Johannes Spalt bereitwilligst und mit grossem Verständnis sein mit den Studenten erarbeitetes, bedeutendes Material zur Verfügung gestellt hätte. Diese Dokumentation ist im weitesten Sinne eine Gemeinschaftsarbeit zwischen dem Leiter der Meisterklasse für Innenarchitektur und Industrieentwurf und den Studenten, deren geistiges Eigentum der Herausgeber übernahm, zusammenstellte und in eine für den Bauschaffenden und Interessierten kurzgefasste Form brachte. Das Ziel der Aufgabenstellung an der Hochschule war, mit dem Aufmessen bestehender Modelle in natürlicher Grösse mit althergebrachten Techniken und bewährten Mechaniken den Studenten vertraut machen und zu eigenen Lösungen inspirieren.

Editor's postscript

The present study, which unfortunately does no more than afford brief insights into the abundance of folding chairs, would not have been possible if my old friend and colleague Johannes Spalt had not been most ready to place at my disposal the important material he had worked up with his students. This documentation is in the broadest sense a work of collaboration between the head of the master class for interior architecture and industrial design and the students, whose intellectual property the editor took over, compiled and arranged in a summarized form for architects and others interested in the subject. The students were given the task at the University of producing full-scale dimensioned drawings of existing models so as to familiarize themselves with traditional techniques and proven mechanisms and to inspire them to find their own answers to design problems.

Meisterklasse für Innenarchitektur und Industrieentwurf o.Ö. Professor Architekt Johannes Spalt, Hochschule für angewandte Kunst in Wien
Klappstühle Naturaufnahmen im Wintersemester 1980/81 wurden von folgenden Studenten gezeichnet und im Massstab 1:7½ (30 cm = 4 cm) wiedergegeben:

Master class for interior architecture and industrial design, Professor Johannes Spalt, University of Applied Arts, Vienna.
During the winter session 1980/81 folding chairs were drawn by the following students and reproduced on a scale of 1 : 7½ (30 cm = 4 cm):

1	Wolfgang Exner	Damenstockhocker	Tripod seat stick
2	Gorgona Staut	Japanischer Hocker	Japanese stool
3	Wolfgang Exner	Jagdstockerl	Shooting stool
4	Franz Schaller	Holzstockerl	Wooden stool
5	Ingrid Lang	Feldstockerl	Camp stool
6	Brigitte Sommersguter	Traghockerl	Portable stool
7	Sabine Mann	Feldstockerl	Camp stool
8	Hannes Huterer	Kinderhocker	Child's stool
9	Hannes Huterer	Kinderklappstuhl	Child's folding chair
10	Marina Hämmerle	Kinderklappstuhl	Child's folding chair
11	Thomas Fichtner	Gartensessel	Garden chair
12	Stephan Ettl	Schiffsessel	Boat chair
13	Stephan Herold	Schiffsessel	Boat chair
14	Stephan Ettl	Englischer Klappstuhl	English folding chair
15	Michael Lang	Kaminsessel	Fireside chair
16	Satria Nandana	Holzsessel	Wooden chair
17	Werner Hollunder	Klappstuhl „Plia"	'Plia' folding chair
18	Johann Mayer	Damenschaukelstuhl	Lady's rocking chair
19	Johann Moosgassner	Verstellbarer Klappstuhl	Adjustable folding chair
20	Peter Miksch	Klapparmstuhl	Folding armchair
21	Eva Schmiederer	Regiesessel	Film director's chair
22	Franz Schaller	Bodensitz	Floor seat

Literaturverzeichnis

Bücher:
Hans Eckstein: Der Stuhl, Die Neue Sammlung München, 1960
Hans Eckstein: Der Stuhl, Keysers Sammelbibliothek, München 1977
Ralph Edwards, A History of the English Chair, Victoria and Albert Museum, London 1951
Ch. P. Fitzgerald, Barbaria Beds, The original of the chair in China, London 1965
H. Kyrielis, Studien zur Formengeschichte alt orientalischer und griechischer Sitz- und Liegemöbel, Berlin 1969
Heinz und Bodo Rasch, Der Stuhl, Akademischer Verlag Stuttgart, 1930
Ferdinand Luthmer, Deutsche Möbel der Vergangenheit, Verlag von Klinkhardt und Biermann, Leipzig, 1913
A. Koeppen, Die Geschichte des Möbels, Bruno Hessling, Berlin, New York, 1904
Adolf Schneck, Neue Möbel, Verlag F. Bruckmann, Stuttgart

Kataloge:
90 a jedna židle, Uměleckoproůmyslove, Muzeum, Praha. 1972
Jochem Jourdan, Ferdinand Kramer Werkkatalog 1923–1974, Schriftenreihe 3 der Architekturkammer Hessen 1975
Rudolf Dirisamer, Österreichische Beiträge zu einem modernen Wohn- und Lebensstil, Zentralsparkasse, Wien, 1978
Gebrüder Thonet, Wien 1904
Thonet Stahlrohrmöbel Wien 1934

Benützte Literatur:
H. Blümner, Das Kunstgewerbe im Altertum, G. Freytag, Leipzig, 1885
Der Schaukelstuhl, Technische Universität München 1979
Erich Klatt, Die Konstruktion alter Möbel, Julius Hoffmann Verlag, Stuttgart 1973
Adolf G. Schneck, Der Stuhl, Band 3 und 4, Julius Hoffmann Verlag Stuttgart 1928, 1927
Stoelen, Delftse Universitaire Pers, Delft 1980
Jürg Uitz, Der Faltstuhl, Dissertation, Graz, 1978

Werner Blaser
Mies van der Rohe – Continuing the
Chicago School of Architecture
2nd edition 1981. 308 pages,
40 illustrations, 70 photos, 12 colour
plates. Hardcover

"This finely produced book from Basel is like a breath of purified air among the gaseous currents of what trendy commentators have already dubbed 'Post Modernism', a term covering a welter of eclecticism (The Good, the Bad, and the Ugly). It is also a reminder of my own ill-fated journey to work under Myron Goldsmith at SOM in Chicago, twelve years ago. What makes it especially interesting is the light thrown on Mies' teaching and influence at the IIT. Examples of student work show that extraordinarily meticulous drawing style nurtured in the early years. It is this narrowing of attention that makes the content of the course appear so thin, but has ultimately produced many buildings of matchless strength and visual refinement..."
(David Wild, RIBA Journal)

Werner Blaser
Courtyard house in China
Tradition and present
Hofhaus in China
Tradition und Gegenwart
1979. 112 pages, 40 photos, 40 illustrations, 4 colour plates. Hardcover
(Text in English and German)

"Werner Blaser, the author of this attractive book, describes the renovated courtyard houses and imperial parks in Peking and the garden complexes in Soochow. The courtyard house, dating back to the Han dynasty, provides an enclosed garden that closes out the outside world. The drawing above of old typical one-story courtyard houses in Peking shows how the houses make the most of scarce land and yet give each occupant the advantages of a garden landscape..."
(Journal of the American Institute of Architecture)

Werner Blaser
Schweizer Holzbrücken
Ponts de Bois en Suisse
Wooden Bridges in Switzerland
1982. Approx. 184 pages, 100 photos,
8 colour plates, 22 illustrations,
20 plans. Hardcover
(Text in English, French and German)

As a result of a wise policy with regard to the preservation of ancient monuments in Switzerland it has been possible to preserve over 150 roofed wooden bridges there. In the neighbouring Alpine countries only about a dozen of these impressive structures have survived. The author of the present book has selected over 30 bridges of this kind, photographed and prepared plans of them. The introductory text gives a comprehensive survey of the history and development of wooden bridge building in Switzerland.

Birkhäuser Verlag
Basel · Boston · Stuttgart